"Not only have I succesfully rehabilitated my injured knee, but I have learned how to prevent further injuries. This method goes beyond stretching; it has changed the quality of my motion."

—John Prather, Triathlete

"Frank Wildman embodies not only the gentleness of this new learning method, he reflects the excitement in the discovery of freedom."

—Carol M. Davis, Associate Professor, University of Miami School of Medicine

"Feldenkrais represents a revolution in human health. Through this method we can learn to improve our living circumstances not only physically but also emotionally, intellectually, and spiritually."

—Smithsonian Magazine

"I had to wait so many years after Moshe to find somebody who would re-awaken my curiosity: someone who would allow me to search for new ways of thinking. This person was Frank Wildman, a new light in the field of the evolution of movement."

—Michel Silice-Feldenkrais, Director, Feldenkrais Institute, Tel Aviv

"The Feldenkrais Method is a discipline for those who look for the glow of their physical and mental life. Frank Wildman hands down and hands on the richness and the complexity of the Method directly from its founder, Moshe Feldenkrais."

—Michael McClure, Obie Award winning Playwright and Poet

"I found this method to be profoundly pleasurable in a way I hadn't expected. After decades of doing yoga, I found I could achieve positions I never had before."

—Hannah Caspari, Yoga Instructor

"Feldenkrais work is the most sophisticated and effective method I have seen for the prevention and reversal of loss of function."

—Margaret Mead

"The system developed by Dr. Feldenkrais has as much potential for understanding the mind/body relationship as Einstein's General Theory of Relativity had for physics."

—Bernard Lake, M.D.

"Frank's Intelligent Body is a great all-around set for both the professional and for general use. The exercises are functionally effective, creative, and a pleasure to do."

—Mark Reese, C.F.T., Ph.D.

Feldenkrais: The *Busy* Person's Guide to Easier Movement

AUDIO AND VIDEO CASSETTES BY DR. WILDMAN

- The Intelligent Body® (Volume I and II)
- The Intelligent Body® (Volume III)
- The Timeless Body: Improving with Age
- The TMJ Tape for Head, Neck, and Jaw Pain
- The Better Driving Tape
- Dealing with Back Pain
- Moving From Pain into Pleasure: Fibromyalgia and Chronic Pain

Available from:

The Movement Studies Institute
P.O. Box 2007, Berkeley, CA 94702-0007
Phone (800) 342-3424
Fax (510) 548-4349
www.movementstudies.com

FELDENKRAIS®

The *Busy* Person's Guide to Easier Movement

50 ways to achieve a healthy, happy, pain-free and intelligent body

FRANK WILDMAN Ph.D., C.F.T.

THE INTELLIGENT BODY PRESS

Berkeley, California

First Printing, 2000

The Intelligent Body Press
P.O.Box 2007
Berkeley, CA 94702-0007

Printed in the United States of America

Cover Art by Lisa Anaya Hubbell
Cover Design by Margret Kaeser
Typesetting by Gavin MacArthur

ISBN 1-889618-75-6
Library of Congress Catalog Card Number 99-95191

Disclaimer:
No part of this book is intended as a substitute for medical advice. In all matters relating to personal health, medical supervision should be sought.

THIS BOOK IS DEDICATED TO THOSE PEOPLE WISE ENOUGH TO REALIZE THAT THEY CAN RECREATE THE WAY THEY USE THEIR BODIES, AND TO THOSE FELDENKRAIS PRACTITIONERS WHO RECOGNIZE THE NEED FOR A PRACTICAL GUIDE TO HELP THEIR CLIENTS.

ACKNOWLEDGMENTS

This book would not have been possible without the persistent efforts of my project coordinator, Rebecca Salome. Her insistence that I produce this book now is largely what made it possible. Her impeccable coordination of the members of the production team was masterful. Some people see the book, some people see the person, and then there is the separate skill of fitting the two together. Hopefully, her inspiration for a series of books to pull out of me over the next few years will come to fruition.

The original German edition of this book, *Feldenkrais: Ubungen fur jeden Tag,* was edited and agented by Vukadin Milojevic and is the basis for this English edition. His talent, insight, and enthusiasm were critical in the development of both books, and I greatly appreciate his participation.

I would like to thank the countless number of students who told me with their bodies exactly what they needed and countless colleagues who have expressed a desire for this kind of book. Being a member of the worldwide community of Feldenkrais teachers has been challenging and rewarding. Their constant input in hundreds of hours of discussion has helped me to shape the lessons included in this book. Many of the lessons have been tried and proven on students of my colleagues, and their feedback helped develop my thinking about how to make this book truly useful to the greatest number of people.

Among friends in the Feldenkrais community, who have helped specifically with this book, is Rosemarie Hausherr, whose good-humored suggestions led to the choice of a title. She also helped us all float through what could have been a difficult process of integrating the outside with the inside of this book. I would also like to thank Paula Batson, Elizabeth Beringer, Sandy Burkhardt, Paul Davidson, and Yvan Joly, for their considered opinions and generous comments. An added dose of gratitude goes to Sanford Rosenberg and Michael McClure, good friends who never stop encouraging me to write.

My thanks to Gavin MacArthur, a typesetter who suffered cheerfully through many changes and worked long hours to make things right, and to Margret Kaeser for her wonderful design of the cover. Added thanks goes to Raymond Marunas for giving valuable advice on the cover and to Anastasi Sioatas, who offered constructive cover suggestions. I also appreciate Rebecca Rhine for so smoothly coping with the additional stresses at the office that publishing this book may have caused.

Finally, a special thanks to my dear friend, and a beautiful painter, Lisa Anaya Hubbell, for her original artwork for the cover of this book.

A Brief History of the Feldenkrais Method®

The method is named after the Israeli scientist Moshe Feldenkrais, D.Sc. (1904-1984). Feldenkrais worked in Paris as a nuclear physicist with the Nobel Laureate Joliot-Curie, developed an electronic antisubmarine detection system, and was the first European to earn a black belt in Judo. After seriously injuring his knee in a soccer game, Feldenkrais learned that surgery had only a 50 percent chance of improving his condition, and would, if unsuccessful, confine him to a wheelchair for the rest of his life.

Unwilling to accept these prospects, he proceeded to learn anatomy, kinesiology, and physiology and combined these with his knowledge of mechanics, physics, electrical engineering, and martial arts (he wrote several books on Judo). These endeavors restored most of the movement to his knee and marked the beginning of a lifelong investigation into human function, development, and learning ultimately leading to the FELDENKRAIS METHOD. Beginning in the 1970s, Dr. Feldenkrais taught his method internationally. He directed the Feldenkrais Institute in Tel Aviv until his death in 1984.

"Through the first years of life, we organize our entire system in a direction which will forever after guide us in that direction. We end up being restricted, we don't do music, we don't do other things. What is more important, we find ourselves capable of doing only those things that we already know."

—Moshe Feldenkrais, San Francisco 1977

Contents

INTRODUCTION

Scores of us undergo unecessary surgeries to restore function to our backs, shoulders, knees, feet, and ankles. We suffer needlessly from pains in the neck and back, irritated joints, sore muscles, and tight jaws. We spend thousands of dollars on pain medication, tranquilizers, or muscle relaxants, and countless hours on boring and ultimately ineffective exercise routines that may leave us feeling as if we were robots or machines. Even worse, millions of us live a shadowy existence in which we suffer quietly and assume that there is nothing we can do about it.

It need not be that way. Imagine yourself deriving pleasure from simple everyday movement. Imagine regaining the mobility you thought was lost forever. Imagine yourself performing activities that you haven't been able to do for years. Now, imagine all of this being possible without pain or effort.

This book will introduce you to a movement method for your brain that will help you to improve your physical and mental abilities in a pleasant and effortless way. You will rediscover the joy and the comfort of easy, well-coordinated movement. You will learn how to use your body's intelligence not only to help you to become a better walker, swimmer, or tennis player, but a better thinker as well.

We spend the first years of our lives discovering how to relate the different parts of our bodies to one another and to our environment. We learn to feel where our arms are, how far our feet are from our head, where up and down is. Remember, human beings are the only animals that must *learn* to move.

Without this learning process we would not be able to use our bodies; we would not know how to move our legs and arms in order to crawl, to keep our balance, or to function in the world.

For most of us this learning process ends once we can move well enough to get by. The playful experimentation with movement and the attentive perceiving and experiencing of what goes on in our bodies stops, and we begin to execute our movements automatically. We become satisfied with performing activities in a habitual and familiar fashion and we cease to refine our movements and to deepen our body awareness as soon as we reach a passable degree of competence. We stop improving how we walk, for example—as long as no one laughs at us.

Our usual movements may serve the immediate and apparent purpose, but we never learn to use the full potential of our bodies; we move with unnecessary effort and overexert certain muscles, while others are hardly ever brought into play. The consequences of this automatic, unaware use are physical complaints, chronic tension, fatigue, and unnecessary deformation that can lead to incapacitation.

We generally assume that all of our problems arise because our bodies aren't able to withstand the exertion caused by a certain activity, or because we're not strong enough and don't have enough stamina, or because the activity itself is too demanding. We try to remedy this situation either by making our bodies stronger and more flexible and increasing our endurance—or by avoiding the activity in question. We rarely consider the possibility that the reason for our complaints is *how* we perform the activity, *how* we move, *how* we use our body.

We deepen our understanding of anything by learning to make distinctions. In order to become a good cook, for example, you must learn to make taste distinctions. You could learn to cook by taking a course or by following instructions in a cookbook step by step. You will actually produce a dish, but you might still feel a sense of insecurity and inadequacy if you didn't use your own sense of taste in deciding which ingredients and in what amount to add. Unless you learn how to make distinctions in tastes, textures, and temperatures, you will never learn how to cook, that is, how to make a good dish under any circumstances.

The movement lessons in this book consist of movements much like a good dish consists of ingredients. They will give you the ability to understand the textures of your body and to develop the necessary awareness to understand what's easy and efficient in a given situation—whether it's typewriting, sitting and reading, making telephone calls, or engaging in a sport. In short, the goal is to use your body's intelligence to move in a more effective way so that you can do all kinds of ordinary activities better.

For more than 20 years I have used the Feldenkrais Method as a basis for teaching people to move with more ease, more comfort, and more efficiency. The method is named after its founder, the Israeli scientist Dr. Moshe Feldenkrais (1904-1984), and forms the basis for the lessons in this book (usually referred to as Awareness Through Movement® lessons). I have worked with people who suffered from chronic pain, as well as with athletes and performing artists. I believe the Feldenkrais Method is the most advanced and effective tool for improving the body's own capacity to learn to move intelligently.

The movements do not require mechanical repetition. Their positive results do not depend on stretching or softening your muscles, but on improving the effectiveness with which the brain coordinates and controls movements. You will have the opportunity to learn how to make use of the unlimited possibilities of your brain—that is why they are called "lessons" and not "exercises."

The lessons are not recipes for "correct" movement, they do not tell you how to breathe or walk, how to sit or stand. They teach you how to become your own measure for efficient movement. You will learn to perceive consciously how you move, where there is tension in your body, where you exert unnecessary effort, and when you are not making use of your full potential. This knowledge will allow you to develop new and effortless movements.

The following lessons will teach you to make very precise distinctions by directing your attention to "small" movements and then combining these small movements into larger, increasingly complex actions. Your ability to sense these fine distinctions, and your growing awareness about how you use your body, will provide you with the basis to move in a secure and relaxed manner—well beyond the immediate scope of the lessons within this book.

I believe everyone, regardless of age, regardless of physical condition, deserves to feel good, to enjoy the wonderful freedom and joy of movement. These lessons are not only designed to help you move more efficiently and healthfully, but they will reconnect you with the pleasure and satisfaction that comes with achieving change.

If you find yourself smiling while you are doing a lesson, you will know that you are doing something right.

—Dr. Frank Wildman

How to Prepare

Before you begin, I would like to offer you a few important tips for getting the most out of the lessons:

1. **Go slowly**
   Time is an extremely valuable tool in the Feldenkrais Method. The movements you are learning may seem unusual and unfamiliar to you. You will need time to assimilate them, to feel the way your body is moving and changing. Do not rush! Pause whenever you feel like it and repeat movements you find pleasurable or want to experience more fully.

2. **Insist on comfort**
   There is no reward in doing any of the movements in an uncomfortable position. Gently alter the position in whatever way makes it comfortable for you. I want you to enjoy the process of the movement as much as the result. If it hurts, it's not helping you (contrary to what you might have been taught). Never try to overcome pain, if you feel it. Pain is a signal that your body is asking you to find a new way to move. Answer it with gentleness and respect.

3. **Don't test your limits**
   The Feldenkrais Method is not about seeing how far you can move, how high you can lift, how long you can stretch. Your goal should be to discover *how* your body achieves a movement so that you can learn to make that movement

easier. Your movements should always be light, and as effortless as possible. Imagine how good it will feel to do simple mobile tasks without trying hard, without working.

4. **Use your imagination**
   Take the time to do different movements from these lessons in your head only, before doing them in practice. Allow the movement to become very clear and lucid in your mind, like a scene from a movie. Imagining a movement before attempting it can make an enormous difference in your ease of motion. You may find that your body responds to your mind by moving as if it is replaying the imagined movement, with almost no effort at all.

5. **Rest frequently**
   The movements in these lessons, while gentle and pleasurable, may cause slight strain because you are using parts of muscles you may not have used in a long time, or in ways that are not familiar to you. Rest often during each lesson. You cannot rest too much. Relax and let the movement settle in, enjoy the feeling. Who knows—it could become a habit.

6. **Choose the comfortable space**
   Learning occurs in direct proportion to the comfort and relaxation your body experiences during the movements. In other words, if you feel comfortable from the beginning of a new lesson, you will be much more likely to learn. For this reason, it is perfectly all right for you to do "lying down" lessons in bed if you find the floor uncomfortable, or if getting down to or up from the floor causes stress or pain.

7. **Don't feel obligated to do all of the lessons**
   Some lessons may be difficult or painful for you to perform, either from the starting position, or at some point during the lesson. If you feel discomfort or pain while moving, stop. Just imagining doing the lesson, or parts of the lesson will benefit you almost as much as if you actually physically performed the movements. Then, as your body

changes and your quality of motion improves, go back to the tricky lesson and explore if and how it has now become possible to perform. If the movements still cause pain, stop and choose a lesson that does not.

8. **Take the lessons with you**
   Throughout your day, pay close attention to how a lesson affected you. One way to do this is to keep a notebook and write down what you have felt from the lesson, and how it influenced the way you performed everyday activites. Be aware of changes in the way you reach, walk, sit, and think. Putting your sensations into words builds a new sensory vocabulary and expands your body awareness, increasing aliveness and changing fixed habits of thinking and feeling. A lesson doesn't have to end with its last movement—let the learning process linger and grow.

Time Key:

The hourglass symbols at the beginning of each lesson indicate the approximate time it takes to complete the lesson—if you go reasonably slowly.

1 hourglass = 5-10 minutes
2 hourglasses = 10-15 minutes
3 hourglasses = 15 or more minutes.

Taking the time to *feel* will create the change. Remember, the slower you go, the faster you will see results!

SECTION I

SITTING IN A CHAIR AT YOUR OFFICE OR HOME

Sitting in a chair is harder on the spine than standing or walking because there is more pressure on the disks of the spine. Many people have to stand frequently because of pressure that accumulates while sitting. Standing relieves muscular contractions that we are unaware of in a static position like sitting. Therefore, if you sit for long periods of time, it is extremely important to have good, stress-free posture with an ability to perform tasks comfortably. The lessons in the following section will help if your job involves sitting at a desk for prolonged periods of time.

For the lessons in this section, please sit in a chair that is flat and firm. An ordinary wooden chair, or a chair with light padding would be best. The flatter the surface the better. The height of the chair should allow your hips to be as high or higher than your knees.

Do not do these lessons in a sofa, armchair, or chair that is low to the ground.

MOVEMENT LESSON 1

THE PELVIC WALK

The purpose of this lesson is to help you discover how to sit more lightly, move your pelvis more easily while sitting, and turn in your chair while performing activities in the workplace or home.

1. Sit comfortably in the middle of your chair without leaning back. Make sure your feet are flat on the floor. Feel yourself sitting as tall as possible. Separate your hands and rest the palm of each hand on your thighs.

✓*Awareness Advice:*

Make sure you don't slump in your chair as you progress through the movements. To heighten your sensation of the movement, you can close your eyes, but remember to maintain a physical attitude of being upright and looking outward.

2. Slide the right side of your pelvis forward in the chair as if you wanted to reach straight ahead with your right knee. Then, slide your pelvis and leg back in the chair. You will be pivoting on your left buttock and sitz bone. Repeat this movement several times until it becomes lighter and more comfortable. Rest briefly and then return to your starting position.

✓ *Awareness Advice:*

Be sure that the work performed in these movements is done with your torso. Keep your feet flat on the floor and do not push too hard with your legs.

3. Repeat the same movement moving the left buttock and thigh forward and backward, pivoting on your right side. Which side glides more easily on your chair?

4. Explore each side again and observe how much your head and shoulders turn. Rest briefly and then return to your starting position.

5. Keeping both feet on the floor, lift the right side of your pelvis off the chair and bring it back down. Do you tilt your whole body to the left, or can you do the movement shortening the right side of your waist and keeping your head approximately in the center?

✓ *Awareness Advice:*

If at any point you cannot feel the movement clearly, or cannot perform it to your satisfaction, stop, close your eyes and *imagine* performing the movement. Then imagine what it would feel like if you were moving and picture the movement happening.

6. Repeat the movement on the other side. Again, which side is easier? Rest.

7. Return to your starting position in the middle of your chair and place your hands on your knees and walk your buttock forward. Then walk the other side forward until you reach the edge of your chair. Then walk your pelvis backward. As you walk your buttocks forward and backward in your chair, make the movement easier.

8. Return to sliding each side of your pelvis forward and backward alternately while looking straight ahead. Work to make the movement easier as you slide one side after the other.

MOVEMENT LESSON 2

CHAIR PLAY

Many people have overly rigid ideas about how they should sit in a chair. Think about all the different ways children move their bodies in relationship to a chair. In this lesson, you will learn to develop more flexible ideas about how to relate to a chair, which will create a much more flexible body.

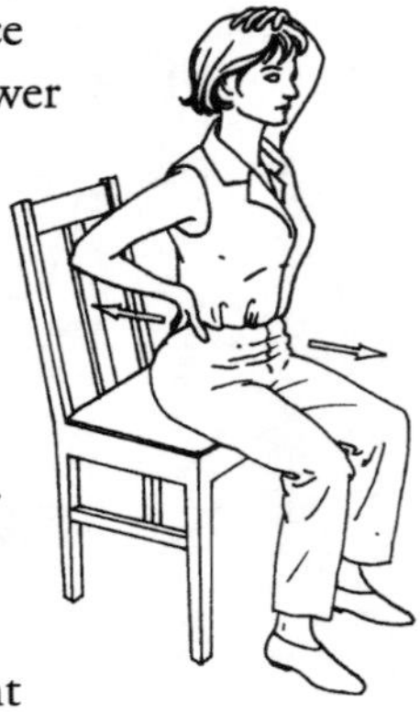

1. Sitting near the front of your chair, place the palm of your right hand on your lower back and the palm of your left hand on the top of your head. Make sure both of your feet are solidly on the floor with your feet and knees well apart.
2. Move your lower back into your right hand by rolling your pelvis on the chair and then roll your pelvis until you feel your back hollow into an arch. As you do this, observe the change in the height level of your head. Rest with your arms down. Are you sitting more on your right side?

✓ ***Awareness Advice:***

If it is too difficult to put the palm of your hand on your back, put the back of your hand there. Your only effort should be in sensing the movement. Use your right fingertips to feel the vertebrae as well as feeling the muscles with your hand.

3. Repeat the same action with the palm of your left hand on your lower back and the palm of your right hand on your head. Which side is easier? Rest. Observe your sitting posture.

4. Sit on the right side of your chair so that the right side of your pelvis is unsupported and only your left side remains on the chair. Put your right hand on your waistline as you would in a casual way with your fingers spread toward your stomach and your thumb in the back.

   Lower and raise the right side of your pelvis so that it goes below the level of the chair and up. Can you feel the right side of your waistline lengthening and shortening? If you put your left hand on top of your head at the same time, can you feel the connection between the movement of your pelvis and your entire spine and neck? Rest sitting back in your chair.

✓ ***Awareness Advice:***

Make sure your feet and legs are fairly wide apart and observe how your right leg assists the right side of your pelvis by pushing the heel into the floor. You might want to experiment by lifting the right heel from the floor as you lower the right side of your pelvis.

5. Repeat the same movement on the other side by having only your right buttock on the chair, with your left hand on your waist and your right hand on the top of your head. Make sure that your legs are wide apart. Is this side more or less fluid than the other side? Rest sitting back in your chair.

6. Sit facing the back of your chair. Lean your folded arms on the top of the back of the chair with your pelvis near the front of the chair. Roll your pelvis forward to hollow your back, and backward to curve it. Try it with your head resting on your arms as well. Can you also rock your pelvis from side to side, pushing through one foot while lifting one side of the pelvis and then the other? Can you do this with your head at rest on your arms as well?

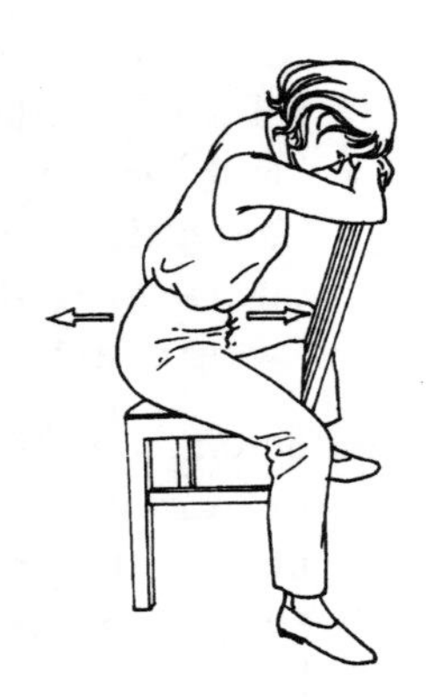

7. Put your hands on your knees and put your chest against the back of the chair with your head looking down and your eyes closed. Can you roll your pelvis forward pushing your belly out toward the back of the chair and then backward, holding it in as your spine curves away from the chair? Let your belly push forward and backward in harmony with the movement of your spine and pelvis until it becomes easy to feel how your breathing can assist the motion. Rest leaning on the back of your chair.

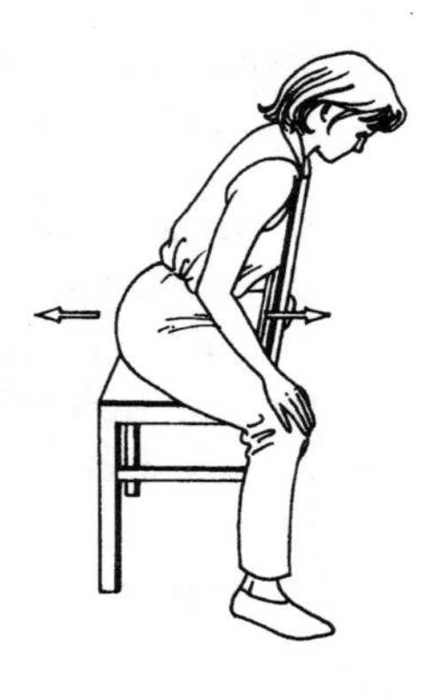

✓ ***Awareness Advice:***

For many people with difficulties in their middle or lower back, it is much easier to sit facing the back of a chair because it provides support and requires the opening of the hips. This lesson or any of the movements in it are useful whenever stress accumulates from sitting.

MOVEMENT LESSON 3

FINDING A MORE ALIVE PELVIS

This lesson will enable you to find how to use your lower back to obtain a more comfortable sitting posture. Many people learn to sit rigidly straight when they are younger, which is too uncomfortable to maintain. Therefore, they eventually find it much easier to slouch. In this lesson you will learn to sit more comfortably and you will learn how to sit taller in an easy, relaxed manner that won't require you to slump for relief.

1. Sit comfortably near the middle of your chair with your feet flat on the floor. Feel yourself sitting as tall as is comfortable. Separate your hands and rest the palms on your thighs or your desk.

2. Imagine that you are sitting on a clock face stamped onto the seat of your chair. Twelve o'clock is in front of you, six o'clock is in back of you, three o'clock is on your right side, and nine o'clock is on your left side. Can you picture all twelve hours of the clock stamped onto the seat of your chair?

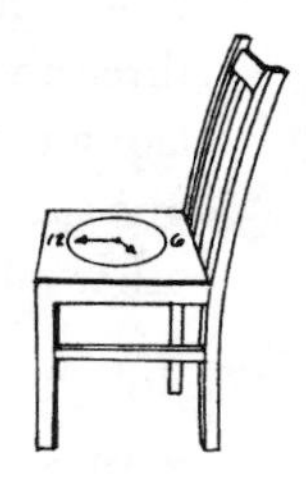

3. Feel the pressure on the bottom of your pelvis in the middle of the clock and roll your pelvis forward to twelve o'clock so that your back arches. Then roll your pelvis backward to six o'clock until you feel your lower back rounding. Are you rolling on the middle of your pelvis, or do you favor one side? As you roll, observe your head moving up and down as you get taller and shorter.

Make the movement as large as is easy, and gradually decrease the size of the movement like a pendulum coming to a halt until you are making a small movement. Then stop in a comfortable place where you can feel your lower back muscles working to support your spine.

✓ ***Awareness Advice:***

As you roll your pelvis forward and backward, keep your chest stationary so that your pelvis and lower back can move independently. Make sure that you are breathing easily.

4. Now roll your pelvis straight across the clock from three to nine. This will feel similar to a movement you did in Lesson 1. Can you keep your head more or less centered? Gradually decrease the size of the movement until you are back to neutral. Rest completely by sitting back in your chair.

5. Roll your pelvis to twelve o'clock and then roll it like a big ball against the rim of the clock between twelve and three and back to twelve. Can you do it several times, feeling where one and two o'clock are? Return to neutral and feel if there is a difference in how you are sitting in your chair.

6. Now, roll your pelvis to three o'clock. Roll it between three and six several times feeling the hours of four and five along the way. Return to the middle of the clock. Does one side of your pelvis and back feel different than the other. Rest if you need to.

7. Roll your pelvis to twelve o'clock and explore the hours between nine and twelve. What does your head do as you explore these hours? Which is easier: twelve to nine, or twelve to three? Return to the middle of your clock and rest. Notice any changes in sensation.

✓ ***Awareness Advice:***

Make sure you are breathing easily while you are concentrating on the movement of the clock. Let your face and mouth be relaxed.

8. Roll your pelvis to six o'clock and explore the hours between nine and six. Is that easier or harder than the hours between six and three? Return to neutral and then take a complete rest sitting back against your chair.

9. Return to the middle of the clock and roll your pelvis clockwise all the way around the clock slowly enough to feel every hour. Return to the middle and rest.

10. Now roll your pelvis counterclockwise. Can you observe which direction is smoother around the clock? Let your upper body move freely as your pelvis adjusts to different hours. Rest in the middle of the clock.

11. Do the original movement from step 3, moving directly from twelve to six. Does it feel more free and easy? Find that midpoint of your clock where your lower back muscles are working to support you and consider this your most efficient sitting position.

✓ ***Awareness Advice:***

Use any of the movements in this lesson to recapture a more comfortable seating position.

MOVEMENT LESSON 4

Freeing Your Middle Back

This lesson will help you to integrate your upper and lower back, as well as your shoulders and neck, into a fuller, more efficient posture. This lesson could also be of benefit to people who experience stiffness or pain in their lower back.

1. Sit comfortably tall in your chair with your feet flat on the floor. Place your right hand on your left shoulder and allow your elbow to hang against your chest. Take your left arm across your chest underneath your right arm, and with your left hand, hold on to your right shoulder. Your right elbow should be resting on top of your left arm. You are now giving yourself a big hug.
2. Keeping your hands on your shoulders, lift your elbows straight up and point them forward, then bring them back down to rest against your chest. Repeat this movement

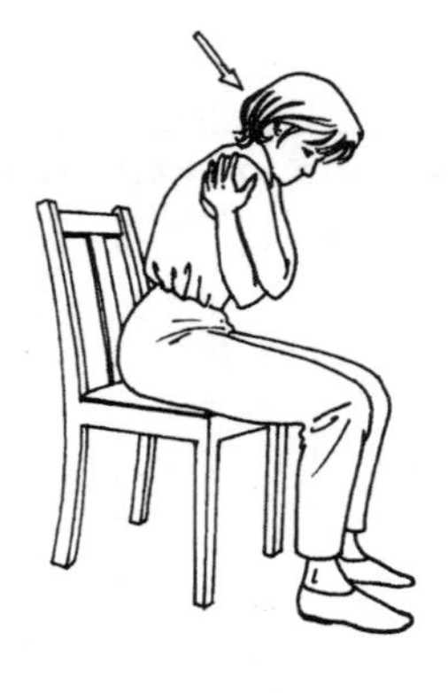

several times, each time allowing your elbows to go higher until you can point at the ceiling, then down to the floor. Look up and down as your elbows are pointing and feel the movement in your upper and middle back, as well as your lower back. Rest in the middle of your chair with your arms in your lap.

✓ ***Awareness Advice:***

As this movement grows larger, you might feel your pelvis rolling forward and backward as it did in Lesson 2. This rolling will help make the movement easier.

3. Repeat steps 1 and 2 with your arms crossed the opposite way. Place your left hand on your right shoulder. Reach with your right hand underneath your left arm and hold on to your left shoulder. Allow your elbows to rest against your chest. As you point your elbows up and down, from ceiling to floor, be sure to let your head follow. Feel how your pelvis can assist you. Rest your hands on your lap or desk and observe any changes in your posture.

4. Cross your arms again as you did in step 1. Raise your elbows to point directly in front of you. Now turn your elbows to your left as if pointing at something on your left side. Look where you are pointing and move your elbows from the center to the left several times. What does your pelvis want to do? Let the right side of your pelvis slide forward and backward in your chair to assist you as you go back and forth. Rest in the middle and re-establish your hands resting on your shoulders.

5. Again, lift your elbows to point in front of you and point them to the right and back again several times. Be sure and allow your head to turn with your arms and let the left side

of your pelvis slide on the chair. Rest completely, leaning back in your chair.

6. Once again, sit in the middle of your chair and cross your arms as in step 3, with the left hand going to the right shoulder first and the right arm underneath the left shoulder. Point your arms forward and begin to turn them right and left several times, while feeling your pelvis pivoting in your chair. Rest your elbows against your chest and re-establish your hands on your shoulders.

7. Lift your elbows again. Keep your nose pointing straight forward and turn your elbows left and right several times without moving your head. Pause. Point your crossed elbows from side to side again and let your head follow. Do you find yourself able to point farther than earlier? Rest completely.

✓ ***Awareness Advice:***

Be sure you are relaxing your jaw and face. Some people have been caught smiling while doing these movements. It's possible. Observe where you prefer to inhale and exhale.

8. Cross your arms again as in step 1. Let your elbows hang down and slowly raise them as you turn to the left so that you find yourself pointing up to your left side. Then bring your elbows down to your chest as you return to the center. Raise them up to the right as you turn to the right. Your elbows will be making a large arc reaching up at the sides and down at the middle. Rest in the middle with your hands on your lap or desk.

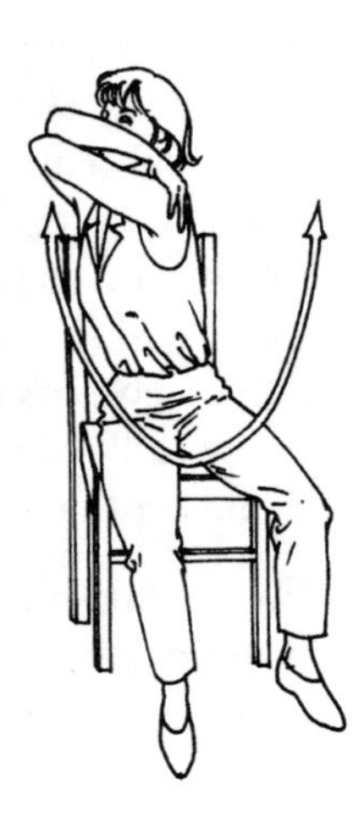

MOVEMENT LESSON 5

FREEING YOUR SHOULDERS

This lesson will help you learn to sit without accumulating stress and tension in your shoulder and neck area. Postural stress in the shoulder and neck is often the result of how we use our lower body. This lesson will work best if you have already done the previous lessons in this section.

1. Sit comfortably tall in your chair with your feet flat on the floor and your hands resting on your desktop, palms down. Close your eyes. Can you feel which shoulder is closer to your ears?

2. Imagine a tiny spider on the tip of your right shoulder that shoots a silken thread to your right earlobe. Sense the length of that thread. Put another spider on your left shoulder and imagine another thread to your left earlobe. Which thread feels longer?

3. With your eyes still closed, raise your right shoulder a few centimeters up toward your ear, and then slowly let it back down. Be sure the movement is slow, and do not move as far as you can go. Rest. Do you feel any change in the sensation of your shoulder?

4. Now lower your shoulder a little bit and let it come up again, and proceed to glide the shoulder up and down, slowly and steadily. Notice the length of the movment of your shoulder up and down. Rest.

5. Now take the right shoulder and glide it directly forward and back to neutral. Rest. Now glide the shoulder backward, about the same amount that you took it forward, and allow it to move forward and backward while observing the length of the movement.

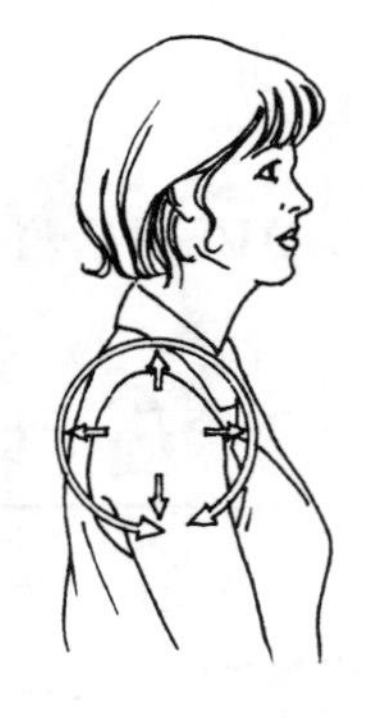

   Which line is longer, up and down, or forward and backward? Can you sense what has happened to your shoulder compared to the other side?

6. Now take your shoulder in a small circle and observe which part of the circle feels smoothest to you. Rest. Repeat again, changing the direction of the circle.

   Sense the height of your shoulder, the length of the silken thread on that side, and feel how relaxed, but awake your shoulder is compared to the other one.

7. Repeat all of the above steps on the other side.

✓ ***Awareness Advice:***

Make sure you are sitting as tall as is comfortably possible, and that your eyes are closed to deepen your ability to feel the movement.

Important! The effect of the lesson will be greater if you make the movements slow, steady, and precise. Fast, sloppy circle movements will not reduce stress.

MOVEMENT LESSON 6

NECK AND SHOULDERS RELEASING INTO LENGTH

This lesson will help you dissolve tension in your neck and shoulders while allowing you to have greater range of motion in your neck. You might become aware of how much the muscles of your lower back and waist, and even the muscles of your legs, are responsible for the feeling of freedom in your neck.

1. Sit comfortably in the middle of your chair away from the back, with your feet flat on the floor. Hold your arms out directly at your sides so that your hands and arms are the height of your shoulders. Face directly forward.

2. Tilt your right ear toward your right shoulder to see how far it goes. Then tilt your left ear toward your left shoulder. What stops you from touching your ears to your shoulders easily? Where does it get uncomfortable? Observe which ear gets closer to which shoulder. Rest your arms down at your sides or on your lap.

✓ ***Awareness Advice:***

As you tilt your head toward the shoulder, make sure your head does not turn. For many people, it is not that easy to tilt their heads while continuing to look directly forward.

3. Raise your right arm directly up to the ceiling until you bring your arm and shoulder to touch your right ear and

the right side of your face. Glide the side of your head up toward your elbow and down toward your shoulder, keeping your head still and facing forward. It's as if you are scratching your arm, up and down, with your cheek. Make only a few movements and then bring the arm down and rest.

4. Raise the right arm, again touching the side of the face to the shoulder and arm. This time keep your head still while you glide your arm higher and lower, as if you were massaging the side of your head with your arm. Rest your arm down at your side and observe the difference in the height of your two shoulders. Has the right one lowered?

5. Hold both arms out at your sides again and tilt your right ear toward your right shoulder. Do you go further? Is it easier? Go to the other side as well. You may find that the left ear gets closer to the left shoulder as well. Rest.

6. Repeat steps 3 and 4 on the left side. Make sure to take frequent rests so that your muscles don't get tired.

7. Raise your right arm to the ceiling again and glue your right shoulder and right ear together so they cannot separate. Now make circles in the air above you with your arm. Use your entire trunk and even your pelvis to make the circles larger. Reach forward and backward and to both sides. Finally, reach with your arm to the right side and hold on to an imaginary paintbrush. Could you write your initials on an imaginary canvas over to your right side? (See previous drawing). Make sure the initials are created with your trunk and that your ear and shoulder are still together. Rest.

✓ ***Awareness Advice:***

While making the circles and writing your initials, make sure your feet are placed wide apart so you can use your legs to support yourself. You might feel your right leg working when you make the initials on the right side.

8. Repeat step 7 on the left side. Rest.

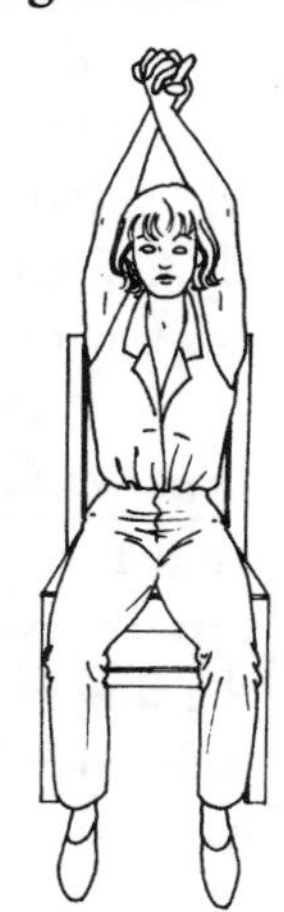

9. Raise both arms up to the ceiling. Cross your wrists and turn the palms of your hands toward each other to interlock your fingers. Can you straighten your elbows so that both arms touch both sides of your head? Can you brush the insides of your arms with the sides of your head by gliding your head forward and backward between your arms?

✓ ***Awareness Advice:***

As your head glides forward and backward between your arms, make sure that there is friction against both arms and both sides of your head. You might feel as if your head is popping out between your arms.

10. Hold your arms out at your sides again and tilt your ears to your shoulders. Can you feel the improvement?

MOVEMENT LESSON 7

SURGEON'S HANDS

This lesson will help you sensitize your hands and allow them to be more mobile and dexterous. If you do this lesson of exploring your hands regularly, you will find improvement in your ability to use your hands for many tasks.

1. Sit in your chair in any way that is comfortable. Place your elbows on your desk or table and very slowly interlace the fingers of your hands. Feel which thumb is on top and which index finger is on top. Then, keeping your eyes closed, slowly unwrap your fingers and interlace them again with the opposite thumb and finger on top.

✓ ***Awareness Advice:***

Your first way of interlacing your hands is almost certainly your habitual way. Reversing the position of the fingers is your nonhabitual way. Can you feel the difference in the familiarity between the two ways? Make sure you do this with your eyes closed.

2. Resting both elbows on the table, touch your right thumb to your left thumb and begin tracing the shape of your left thumb from its base to the tip on all sides of the thumb. Which thumb feels more sensitive?

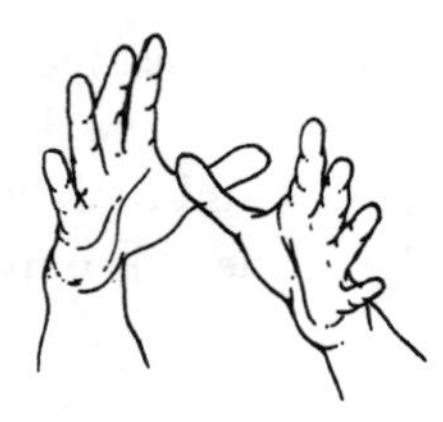

Rest with your hands on the table and feel the difference in the sensation of your two thumbs.

3. Hold your hands up again, touching your right index finger to the tip of your left, and begin tracing all sides of the left index finger from base to tip. Feel the space between the index finger and thumb, as well as the space on the other side. Can you feel all the joints? Can you feel the difference between the active finger and the passive finger? Rest your hands.

✓ ***Awareness Advice:***

As you proceed through this lesson, make sure to move very slowly with the intention of feeling all parts of your fingers.

4. Now touch the middle finger of your right hand to the tip of your left middle finger. Again, explore the entire finger and the spaces on each side. Usually, the middle finger is the most sensitive on the hand.

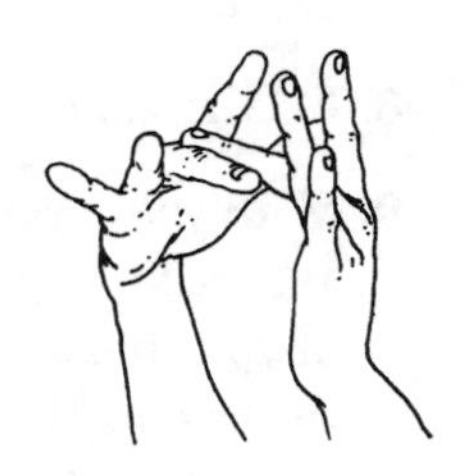

5. Do the same thing with the fourth fingers, and little fingers. Rest your hands on the table with the palms down and feel the difference in the sensation of the two hands. Explore opening and closing the left hand and feel the clarity of the movement of your fingers and hand. Try the same with your right hand to feel the difference. Notice how the changes are reflected in the two sides of your face.

✓ ***Awareness Advice:***

This lesson will be effective only if you move slowly and sensitively. It could take several minutes to explore all five fingers.

6. Bend your elbows so that once again your hands are in front of your face. Close your eyes. Take your left thumb to touch the tip of your right thumb and sensitively explore the shape of your left thumb. Then proceed, finger by finger, to explore your right hand's fingers with your left hand's fingers as you did in steps 2 through 5. Rest and explore the sensation and movement of your right hand.

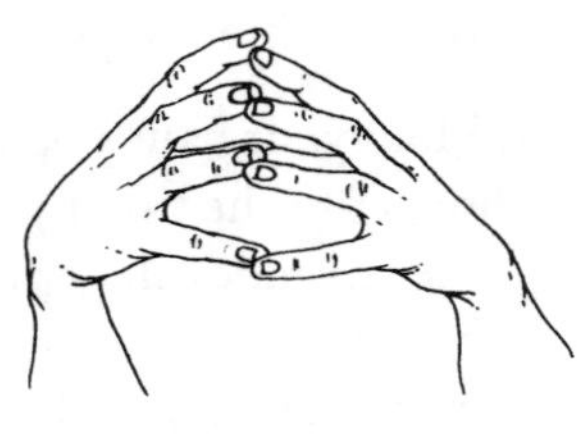

7. Close your eyes and bring your left index finger to gently touch your left eye. Can you feel the shape of your eyeball? Do the same movement with all the fingers of your left hand. Rest.

8. Do the same movement on the right side.

9. With your eyes closed, touch all your fingertips together and make tiny circles at your fingertips. Gradually slide all of your fingertips together, interlaced as at the beginning of the lesson. Slide them out and re-lace them the opposite way. Does this feel different? Throughout the day, observe how your hands perform familiar tasks.

MOVEMENT LESSON 8

SAFE BENDING WHILE SITTING

Many people suffer the most severe injuries to their back doing simple things such as bending down to pick up a dinner napkin or a pen that they've dropped on the floor. When they go to straighten up, the muscles don't know how to contract. This lesson is a wonderful preventive measure for back strain, and it just might improve your sitting posture as well.

1. Sit in the middle of your chair with both feet flat on the floor and the palms of your hands on your thighs, near your knees. Place your feet just slightly in front of the knees.

2. Slide your left hand down the outside of your left leg toward your foot, only as far as it's very easy to go. Massage back and forth down toward your foot, back up to your knee, and along the side of your leg to your hip. Massage slowly and carefully several times until the movement feels clear and easy. Rest and feel the difference between the two sides of your body.

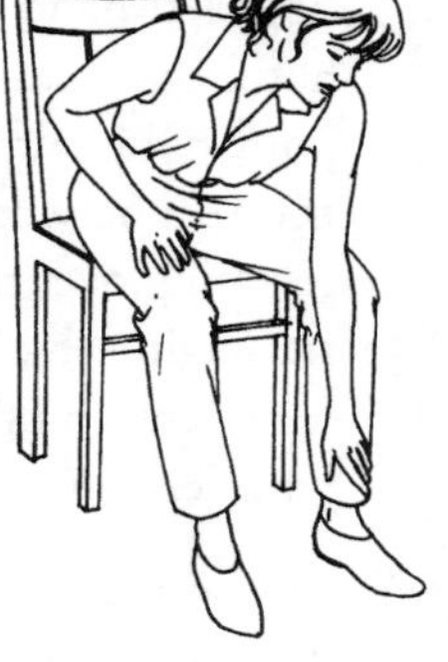

3. Do the same massaging motion with your right hand, over your knee, toward your ankle and then up over the hip. Do it as thoroughly as you did with the left hand. Rest.

✓ ***Awareness Advice:***

Be sure your feet are fairly wide apart, both for this massaging action and throughout the lesson, as it will help stabilize you and give you better balance. You wouldn't want to fall out of you chair, would you?

4. Now take your right hand and put it on your left knee and begin to massage down toward the foot and back again. Rest.

✓ ***Awareness Advice:***

There are two ways to do this. You can either push with your left foot into the floor as your hand goes down and up or you can rest your left hand solidly on the left thigh and use the muscles of the left arm to assist you as well. If your back bothers you, use both the left hand and the left foot for support.

5. Put your left hand on your right knee and massage down toward the foot and back again, as you did on the other side. Rest.

6. Now take your left ankle and place it over your right knee and take your left hand between your legs and begin massaging the back of your right calf, up and down with the arm between your legs. Try reaching over the left leg and massaging the front of your right leg a few times as well. Uncross your legs and rest.

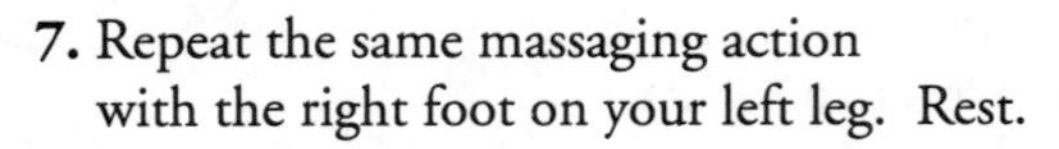

7. Repeat the same massaging action with the right foot on your left leg. Rest.

8. Now stand your right heel on the front edge of your chair and massage your left leg with your left hand. Massage both sides of the left leg, toward the foot and away. Rest.

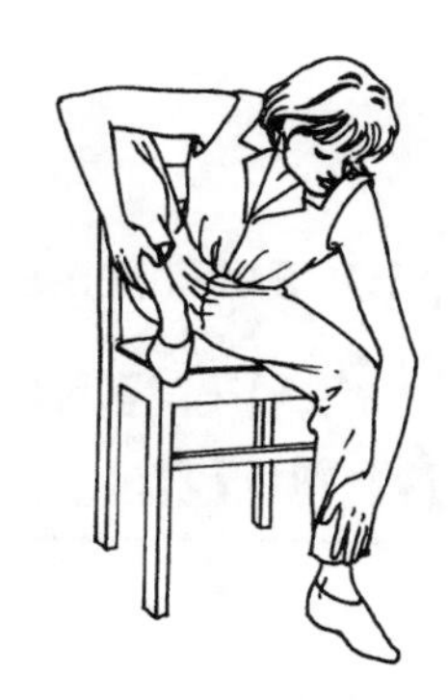

9. Repeat the same movement on the other side with your left foot on the edge of the chair, while massaging your right leg with your right hand. Rest.

10. Now sit toward the front of your chair and bend your knees enough so that only your toes remain on the floor. Can you slide your left hand down toward your left heel, or perhaps even further? Go up and down several times. Rest.

11. Try the same movement with the right hand reaching toward the right heel and foot, on the right side.

12. Again, sit with both feet wide apart on the floor, both palms on your knees. Now explore sliding both hands down the front of your legs, toward the floor and back again. Rest in sitting. You may find it easier to sit in your chair now.

✓ ***Awareness Advice:***

Repeating any part of this lesson will help you to avoid major difficulties that lead to sometimes serious back pain.

MOVEMENT LESSON 9

Preventing Repetitive Stress Injury: Mobilizing Neck, Shoulders, and Back

This lesson offers an easy and fast way to relieve yourself from the stress that accumulates in your neck while sitting for prolonged periods of time. Overuse syndrome in the wrists and fingers comes from too much friction on the tendons in those areas. When we work with our hands, there must be a response, like an echo of the movement, through our shoulders and into our trunk. If the shoulders and neck are stiff and this response to the movement of our hands can not occur, the only place left for motion to take place is in the hands and wrists alone. In other words when you move your hands, it is more efficient to have a mechanical response in your whole body.

1. Sit at your desk, with your chair pulled away far enough so you can lean forward with your arms and head on your desk to support you from falling forward. Your feet are firmly on the floor to support you and your body is at approximately a 45- degree angle to the floor. Place your arms on the desk with your elbows pointing outward and your fingers touching above your head.

✓ ***Awareness Advice:***

Make sure that you can breathe comfortably in this position and that your feet are not too close to your chair. Practice resting both sides of your face, as well as your forehead, on the desk until you feel comfortable. If your desk feels too hard, place your head on the backs of your hands.

2. With your forehead down on your desk, or on the backs of your hands, lift your head to look up toward the ceiling slowly and easily, keeping your elbows in place. Observe how far you go without effort. Can you feel the movement in your back?

3. Keeping your hands near each other, and the elbows wide apart, put your forehead near the edge of your desk and reach forward with your chin to touch your fingertips. Go back and forth several times. Alternate touching your chin to different parts of your hands and wrists. Rest by placing your forehead on your desk near the edge. (You might want to place a towel or something soft under your forehead.)

4. Slowly roll your head from side to side by letting your shoulders and your torso assist in the movement.

5. Turn your face to the left and place your right cheek on the backs of your hands. Slowly glide your face toward your left elbow as if your nose were aiming for the inside of your elbow. Then slide the back of your head toward the right elbow. Glide your head several times to the left and right while looking to the left. Rest sitting upright.

✓ ***Awareness Advice:***

Make sure that your shoulders move as your head glides. Can you feel your shoulders alternately moving up and down?

6. Place your cheek on your hands looking to the right and glide your nose toward the inside of your right elbow, and the back of your head toward the left. Each time you glide your face toward your elbow, use a different part of your face. Try your chin, mouth, forehead, etc. Rest sitting upright. Observe how you sit. Are you more upright? Take a walk through the room and observe if this lesson has affected your standing posture.

MOVEMENT LESSON 10

Integrating Your Neck, Shoulders, and Back: Hands to Head

This lesson will help you gain more flexibility in your neck and upper back. Learning to mobilize the shoulders, upper back, and neck will alleviate stress not only in those areas, but also in your arms and hands. Steps 1 and 2 are repeats from the previous lesson.

1. Sit at your desk, with your chair pulled away far enough so you can lean forward with your arms and head on your desk to support you from falling forward. Your feet are firmly on the floor to support you and your body is at approximately a 45-degree angle to the floor. Place your arms on the desk with your elbows pointing outward and your fingers touching above your head.

✓ ***Awareness Advice:***

Make sure that you can breathe comfortably in this position and that your feet are not too close to your chair. Practice resting both sides of your face and your forehead on the desk until you feel comfortable. If your desk feels too hard, place your head on the backs of your hands.

2. With your forehead down on your desk or the backs of your hands, lift your head to look up toward the ceiling slowly and easily, keeping your elbows in place. Observe how far you go without effort. Can you feel the movement in your back?

3. Now place your right hand on the back of your left. Turn your face to the right to rest your left cheek on the back of your right hand. Imagine that the back of your hand and cheek are fastened together with glue. Could you lift your right arm and head together as a unit so that your elbow, hand, and head all rise together at once? Do this movement several times until it feels clear and easy.

   The next time you lift your head, make a circle in the air with head and your arm so the top of the circle reaches up toward the ceiling and the bottom is at your desk. Make the circle round and large. Rest and observe how you are sitting up now.

4. This time, keeping your right elbow on the desk, lift your right hand and head together as a unit several times. Can you feel the movement in your ribs? Rest.

✓ ***Awareness Advice:***

Remember that the back of the hand and the cheek should move as if fastened together so that all movement takes place in your upper and middle back. Can you feel the difference between moving the entire arm with the head and keeping the elbow on the table? Each way will mobilize different parts of your back.

5. Place your left hand on the back of your right hand and look to the left with your right cheek resting on the back of your left hand. Repeat steps 3 and 4. Which side is easier? Rest leaning back in your chair.

6. Lean on your desk in the original position of step 1. Turn your face to the right and place your left cheek on the back of your left hand. Take your right hand and place it on the right side of your head. Proceed to lift your head and right hand as one unit, pivoting on your elbow. Do this movement until it feels easy and familiar. Rest.

7. The next time you lift your right hand and head, can you go up until your right arm is vertical? (Some people sit at their desks this way, anyway.) Now sweep your right elbow across your desk to your left hand and then up your left arm as far as is comfortable. Then set your head back down on your left hand. Each time you lift your head to the vertical position, sweep your right elbow a little further to the left. (Can you see over your left shoulder?) Sweep all the way back again to rest your head on your left hand. Rest sitting upright in your chair.

✓ ***Awareness Advice:***

Each different position of the hand on your head creates the possibility for a different part of your shoulders, ribs, and spine to become mobilized. By constraining the ordinary way you might move your head, your brain will find a new place to move to fulfill the instruction. So, be precise.

8. Repeat on the other side by placing your right cheek on the back of your right hand and your left hand on the right side of your head. Rest and observe your seated posture.
9. To conclude, place both hands on the desk again with your elbows wide and your fingertips touching. Rest your head on your hands or the desk and look up to the ceiling again a few times. Does your back participate more easily? Do you find your range of motion has improved? Do you find your chest and shoulders are more comfortable?

MOVEMENT LESSON 11

Rising from Sitting to Standing

In order to stand and sit easily, the freedom to move parts of your spine and pelvis is a basic requirement, especially if you have difficulty with the hips, knees, or ankles. Lessons 1 through 4 will increase the effectiveness of this lesson. Please be sure to learn them first.

1. Sit close to the front of your chair by sliding or walking your pelvis forward. Establish a position where your left foot is slightly behind your right foot. Your feet and knees are spread as far apart as the width of your pelvis. To make sure that your knees and legs are relaxed and free, can you flip-flop your knees back and forth until there is no tension when you stop? Rest, feeling tall in your chair with your focus out into the room.

2. Keep your spine straight and think of your trunk as a single unit. Could you begin to rock your body forward and backward until your head moves in a large arc and you feel your pelvis rocking on the chair? As you rock backward, try to create a motion that mirrors the forward arc. Can you feel your right leg helping to push you back as you bow forward?

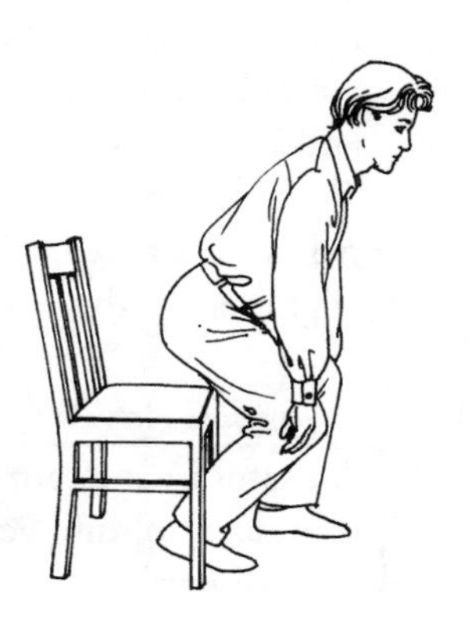

3. With your hands on your knees, sitting close to the edge of your chair, rock forward until you feel enough weight over the forward foot that you can stand by simply straightening your knees. You should be standing with your left foot behind your right foot.

✓ *Awareness Advice:*

The forward foot should be placed directly below the right knee, but you will have to explore precisely where it's easiest for you. If you have pain in one of your knees, you might prefer practicing with the forward and backward legs reversed immediately, until you learn how to rise easily. Although you are rocking your weight toward the forward foot, make absolutely sure that your are pushing through both feet to come up to standing.

4. Put the palms of your hands in the crease of your hip, where your thigh meets your trunk. Very slowly bend your knees as you bow forward. Feel a deepening in the crease of your hip. Keep bowing with your head reaching forward until your pelvis "discovers" the front of your chair. Practice getting up and down this way several times. There are two points of focus at this point—where your pelvis meets the chair and where your head is directed in alignment with your trunk.

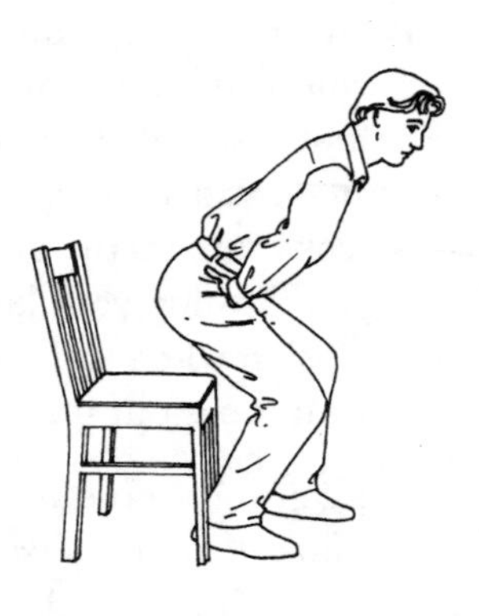

✓ *Awareness Advice:*

A useful image to make standing up easier is of someone holding your head and gently pulling you forward.

5. Explore sitting to standing and back again, with your left foot in front of your right foot. Rest.

✓ *Awareness Advice:*

If the movement up from the chair and back again is to become easy, it must be completely reversible at all times. This means that if photographs were taken of the sitting and standing motions, no one would be able to tell if you were moving toward sitting or standing positions. The trajectory of the head, chest, and pelvis is identical in either direction.

6. As you move up to standing with one foot behind the other, allow yourself to turn to the side of the backward foot, as if you wanted to stand and look to one side. To sit, reverse the same spiral. Practice this on both sides, changing the backward foot.

7. Now, place both feet apart directly under your knees. Sit on the edge of your chair. Rock forward as if your eyes and mouth are reaching for something and see how effortlessly you can stand. To sit, bend your knees and reach your eyes and mouth forward for something, keeping your breath free until your pelvis finds the chair.

✓ *Awareness Advice:*

We rarely stand for no reason. It is much easier to stand when we have the intention of going somewhere. For example, we stand to walk out a door, or to look at something, or to get somewhere to perform a task. So, as you practice sitting to standing, think of orienting yourself toward something in the room, and let that place pull you out of your chair.

8. Practice getting up and down with the feet directly under your knees by turning to look slightly behind yourself as you stand. Alternate from side to side, feeling yourself spiraling up and down easily. Look straight forward and come to a standing position with the intention of walking in the direction you are looking. Remember to make the trajectory of the movement reversible when you sit again. Rest. The next time you are standing, observe if walking seems lighter.

MOVEMENT LESSON 12

Reaching with Ease

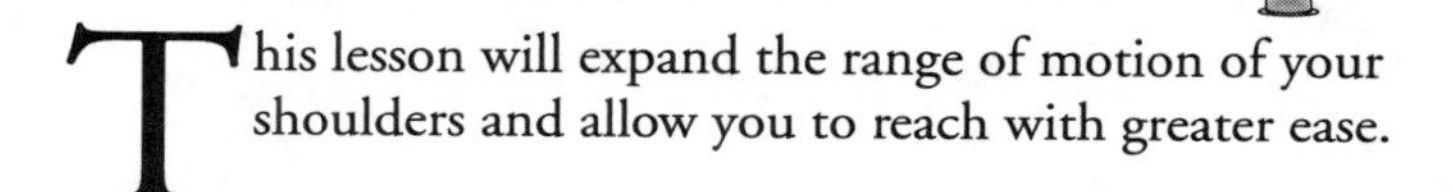

This lesson will expand the range of motion of your shoulders and allow you to reach with greater ease.

1. Stand in front of a wall with your toes only a few inches or centimeters from the base. Put your right hand on the wall, at about the height of your shoulders.
2. Slowly slide your hand up toward the ceiling, and back down. Do this several times. Rest while standing.
3. Repeat with your left hand and arm. Rest.
4. Stand with your right side against the wall. Put your feet apart so that the right side of your foot touches the wall. Put your right arm straight out in front of you with your palm against the wall, and slide the hand forward and backward on the wall. Your shoulder will also slide forward and backward. Allow your body to turn as the arm glides back and forth. Rest and walk away.

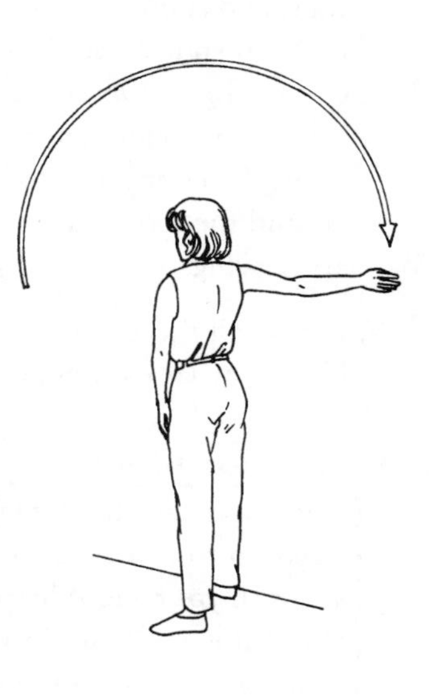

5. Go back to the wall, and again place the ankle, hip, and shoulder against the wall. Place the hand in front of you again. Slide your hand up the wall until it reaches above you. Slide it all the way in back of yourself, looking at your hand while you do the movement. When the hand is in back of you, can you touch your shoulder against the wall? Now bend your right elbow and slide the back of your hand against the wall so that your hand can slip under your waist, in the opening between your hip and shoulder, until it is in front of you again. Go around in large circles this way several times.

✓*Awareness Advice:*

It's important to keep your right foot against the wall, as well as your hip and shoulder. It's also important to allow your head to freely observe your hand's circle. If you cannot keep your right shoulder against the wall the whole time, at least try to keep it close.

6. Take the right arm in circles again in the opposite direction. Begin by aiming your elbow toward the gap between the waist and shoulder, allowing the back of your hand to slide through your waist. Get the feeling of making large, lazy circles. Rest. Observe the difference in the feeling of each side of your shoulder, and the feeling of the length of your arms. Does one arm feel longer? Go for a short walk around the room and feel the difference in the way your arms swing.

7. Repeat all the above movements on the left side. Rest and observe what has happened.

✓*Awareness Advice:*

It is important to be playful and to allow your body freedom to adjust as you need to so that there is no stress in your shoulder joints. However, being precise about the position of the foot, hip, and shoulder will create an expanded range of motion for you.

8. Stand facing very close to the wall again, and rest your toes and chest against the wall. Place your hands against the wall, beneath your shoulders, as if you were going to push yourself away. Slide both hands up, as if reaching for the ceiling until your elbows and shoulders are straight. Try doing this with your forehead against the wall as well. When your arms are high above you, explore sweeping your hands along the wall. Can you feel your shoulder blades responding? Rest and go for a walk around the room. Observe how open your chest feels, and how wide your shoulders feel.

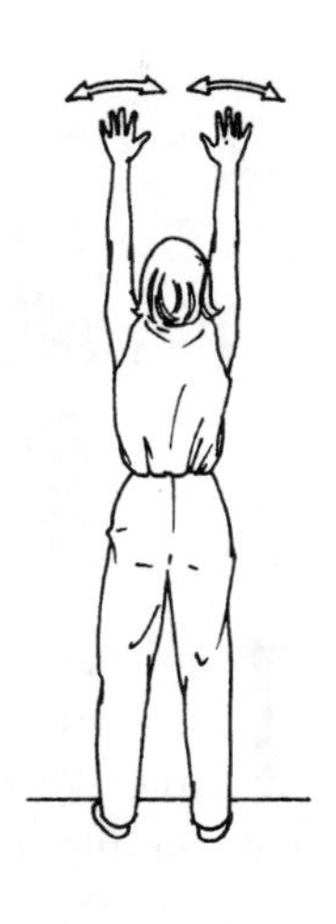

SECTION II

Reducing Tension in Your Jaw, Mouth, Face, and Neck

In these lessons, you will learn to improve the function of your mouth and jaw, so movement can be more comfortable. You will also learn to reduce the muscular stress and tension in the jaw, mouth, face, and neck.

The head, neck, and jaw are common sites for pain. Some difficulties require you to see your dentist or oral surgeon. But you can supplement any program with these exercises and have added relief.

You will be doing gentle movements involving the eyes, face, and tongue, integrated with the jaw. Stress in any of these areas can affect the jaw. The lessons in this section will allow you to use your jaw more easily by breaking up rigid and painful patterns of work in all the muscles of the mouth, jaw, face, and neck.

Many people find that these lessons improve the expressivity of their face, make it easier to breathe more comfortably, and improve their voice.

You can do these lessons sitting or lying on the floor. If you have difficulties with any of the lessons, learn to do them both sitting and lying. The lessons will be described in the seated position.

Do all the movements slowly and with attention. The effectiveness of these lessons depends upon how slowly and attentively you do them—not on the number of repetitions. The slower you do the movements, the easier they will become. Having a sense of ease and comfort are the most important things in doing these lessons. If you feel pain, make the movements smaller and produce the movement with less effort.

MOVEMENT LESSON 1

STRESS-FREE JAW

This lesson will allow you to move your jaw more easily and comfortably. If you have pain or difficulties with your jaw, you might find that the unusual nature of the movements in this lesson will help increase the ease of movement as well as the range of motion. Remember, if you feel pain or discomfort, move more slowly and decrease the size of your movements. In the Feldenkrais Method, the size of any movement will increase by learning to be comfortable in a small range that gradually expands. If we force things to move too quickly, they can become irritating and we lose what we have gained. So remember, less is more.

1. Sit in a comfortable chair at your desk or table. Close your eyes. Notice the position of your tongue—where is it? Can you feel whether or not your chin is directly in the middle of your face or off to one side?

2. Slowly open your mouth a few times to notice how much effort is involved. If you close your mouth slowly enough, notice if your teeth touch first on one side. Many people never realize the asymmetry in the mechanics of their jaw.

3. Sitting upright, lean forward to rest your elbows on your desk or table. Take the fingers of both of your hands, and hold on to your lower jaw as in the illustration.

Can you work your thumbs along the bone underneath the chin? If you give a gentle pinch between your fingers and your thumb, can you feel your lower jawbone? While holding your jaw in place with your hands, open and close your jaw by moving the skull up and down. Rest.

✓ ***Awareness Advice:***

In this action, your jaw will open because your skull is going up and down, which reverses the usual relationship between the jaw and the skull. A basic principle—moving a large mass against a smaller will create a greater and easier range of motion. Sometimes people with severe jaw pain shift the work load from the muscles of the jaw to the muscles in the back of their neck, and open their jaw by compensating with their neck.

Go slowly and not nearly as far as you can.

4. Once again, hold your jaw in your fingers on each side. Keep your jaw stable while carefully turning your head from side to side. Can you feel your eyes turning with your skull? Pause briefly, and then, again, glide your jaw from side to side with your fingers assisting. Rest.

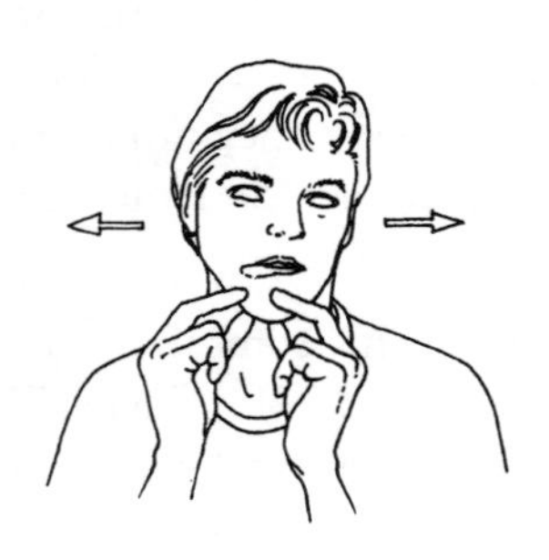

5. Let your tongue come out of your mouth and slide it over your lower lip from side to side. Allow your jaw to follow your tongue. Let the movement grow so that your head turns a little to each side as it follows your tongue and jaw. Continue turning your head to each side, leading with the movements of your jaw. Rest.

✓ ***Awareness Advice:***

Repeat all of these movements occasionally to break the rigid habits held in the muscles of your jaw.

MOVEMENT LESSON 2

The Big Mouth Lesson

This lesson will expand the sensation in your mouth and free the motion of your jaw. Many people find that this lesson gives them an image of a very large mouth, which is both unusual and enjoyable.

1. Sit upright and comfortably in a chair or lie on the floor with knees bent and feet flat on the ground. With your eyes closed, observe the position of your jaw and tongue.

2. Place your tongue in the space behind your upper front two teeth. Move your tongue over both the back and front surface of each tooth on the right side of the upper row. From the midline to the right side, count the number of teeth with your tongue. Rest. Sense the size of your mouth and your jaw. Has your jaw changed position?

3. Repeating the exercise with the teeth of your lower jaw, place your tongue in the space behind the middle front two teeth. Move your tongue over both the back and front surface of each tooth on the right side of the lower row. From the midline to the right side, again, count the number of teeth with your tongue. Rest. Sense your mouth and jaw. Is there a difference between the two sides? Which side do you prefer?

4. Now make a circle with your tongue against the inside of your right cheek. Make the circle slowly, evenly, and precisely. Can you make the circle in both directions?

What are your eyes doing? How are you breathing? To help clarify the circles, put your hand on the outside of the cheek and feel the tongue pushing on the hand. Rest.

✓ ***Awareness Advice:***

In performing precise movements of the tongue and jaw, you might find the rest of your body becoming tense. Some people hold their breath and stop moving when they finally run out of oxygen. Although you need to concentrate on what you are doing, let your attention drift through your body to make sure that you are not overworking your shoulders, hands, or feet.

5. Now put your tongue inside the circle and write your initials. Again, use your hand against the cheek if you wish. Rest.

6. Put your tongue in the midline over your upper teeth in the groove between the gum and lip. Run your tongue in that groove all the way to your circle and return to your midline in the groove in front of your lower teeth. Sweep your tongue deep in the groove from upper to lower teeth passing the inside of the cheek. Can you feel all the movements of your jaw, neck, and eyes? Rest.

Observe the difference in the sensation between the two sides of your face and mouth. Has your jaw changed its resting position as well as your tongue? Does the right side of your face feel different as well? Slowly turn your head from side to side. Is there a favorite side? You might want to stand up and walk a little bit and observe to which side it is easier to turn your body.

7. Repeat this entire lesson on your left side.

MOVEMENT LESSON 3

EXPANDING YOUR THROAT

Many people do not realize that they squeeze their throats all day long, affecting their voices, the ease of their breathing, and their ability to swallow comfortably. This lesson will help you control and expand the size of your throat and release deeply held tensions inside the mouth and around the voice box.

1. This lesson could be done sitting comfortably in a chair, or lying on your back on a carpeted floor. Take a moment to sense the space inside your mouth. Can you feel or imagine the space all the way down through your throat and into your lungs? Into your stomach?

2. Make the space inside of your throat and mouth smaller by slowly squeezing with the muscles of your throat and mouth. Can you feel the tension spread into your neck, shoulders, and face?

3. Now, keeping your lips together, could you expand all that space in the mouth and throat? Alternate between expanding and collapsing the space inside.

✓ ***Awareness Advice:***

Our throats and mouths expand naturally whenever we yawn. If you're not sure how to expand your throat, keep your lips together and create the beginning of a yawn. If it's still not clear, open your mouth wide and yawn, feeling what happens in your throat and to your jaw. Then, try yawning again with your lips together.

4. With the space in your throat and mouth collapsed, say "Hello" and then your name, and hear what it sounds like. Also, make some vowel sounds like "Oh" and "Ah" and hear their quality. Rest.

5. Now expand your throat and repeat the same sounds or words. Can you hear the difference? And can you feel the difference in your chest, shoulders, and neck, as well as your jaw and face?

6. Alternate between expanding and constricting your throat, and experiment with different sounds until you feel you have control and choice over the habitual tightness that constricts the area.

✓ ***Awareness Advice:***

Whether you begin this lesson lying down or seated in a chair, be sure to explore the actions in this lesson in both positions. Gravity passes through our muscles at a different angle when lying or sitting. You will deepen your control if you can learn to reproduce the movement in both positions.

MOVEMENT LESSON 4

DEVELOPING A MORE EXPRESSIVE FACE

This lesson will reveal how the muscles of the face act in harmony with the actions of the jaw. If the muscles of the lips, the cheeks, and the rest of our face do not move, it is more difficult for the jaw to move freely.

This lesson might be more easily learned lying on the floor on your back, then practiced sitting upright in a chair.

1. Put both index and middle fingers just below your ear and find the hinge of the jaw. With your eyes closed, open and close the jaw slightly to help find the hinge. Slide the jaw side to side as well. Can you feel it in your fingers?

2. Can you glide your jaw forward and backward? Do you feel more strain in the throat or neck while doing this movement? Can your lips facilitate the gliding movement?

3. Stop the movement of the jaw, and take your lips forward and back. How does that affect your jaw? Reach your lips forward in a round "O" shape as if you were reaching your lips to kiss someone. Can you feel your jaw glide forward automatically?

4. Now, slowly pull the lips back and take them so far back that you find yourself with a big grin on your face. If you slowly smile, you will feel your jaw retract. The bigger the smile, the greater the motion of the jaw. Can you feel the jaw gliding backward?

5. Alternate between reaching your lips forward in a big kiss and back in a big wide smile feeling your jaw gliding forward and backward with the movements of your lips. Does your neck participate in the forward and backward motion?

✓ ***Awareness Advice:***

In the seated position, allow the head to glide forward with the jaw as you reach the lips forward, and then, let the back of the head move backward as you move the lips and jaw back into a grin. This movement of the face and head can unlock the jaw for many people who have difficulties.

6. Learn to perform the movement of the lips, jaw, and head fairly lightly and quickly after you have mastered it by going very slowly. Also, try turning your head to either side and perform the same movement while aiming in a different direction. Rest. Notice where your head sits on top of your spine. When you stand up and walk, notice if there is a response of your head to the movement of your legs. If not, can you create it?

✓ ***Awareness Advice:***

The forward and backward action of the jaw is primarily related to the lip reaching and smiling functions of the face. All primates need the social gestures of the face and jaw to survive in their community, even though these actions are not necessary to chew food. You could say that the forward and backward motion of the jaw and face are social movements. The forward and backward action of the head in harmony with this is also natural and is present in the walking of most birds and mammals, most visibly in the chicken and the horse.

SECTION III

Better Driving Lessons

This series of lessons will increase your driving comfort and pleasure. In addition, you will learn to improve your driving position so you will be more alert and responsive behind the wheel. All of the lessons in Section 1, *Sitting in a Chair at Your Office or Home*, especially Movement Lessons 1 through 5, can be done inside the car.

The exercises in this section should be done while parked in your car by the side of the road. You will learn to use your eyes, hands, legs, and back differently than might be familiar to you. So, please do not attempt to learn the lessons in this section while you are under the stress of driving. Also, it is best not to do the lessons in a confined area, like a garage. If you are parked by the side of the road, it is the closest approximation to actual driving. It will be easier to transfer the learning from these lessons into a driving situation. Eventually, you will no longer need to rehearse these lessons and will be able to utilize them while driving.

Always fasten your seat belt while learning these lessons, whether driving or not.

MOVEMENT LESSON 1

Finding Your Optimal Driving Position

Most people don't think much about their driving position, yet it contributes greatly to our reaction time as well as to our pain and comfort. A good driving position will increase your alertness and alertness will increase your driving skill.

So first let's work on your seating position. Regardless of the car seat you have, you can improve your seating position to improve your driving and make it more enjoyable.

1. Sit far enough back in your car seat so that the back of your pelvis touches the back of the car seat. Notice where you feel the greatest pressure. When you are sitting this far back in your seat can you easily reach the brake and gas pedals? Are your elbows straight or slightly bent and pointing down?

2. Readjust your seat so that your hands can fall on the wheel at ten and two o'clock with your elbows hanging down. See that your pelvis and lower back remain touching the back of your seat and that you can easily and quickly reach both the brake and gas pedal without effort.

3. While keeping your shoulders and upper back at rest against the car seat, slowly arch your lower back away from the seat. You will do this by rolling the top of your pelvis forward. Now roll the other direction and push your lower back into the car seat. Go back and forth several times. (If you find this difficult, go to Lessons 2 and 3.)

✓ ***Awareness Advice:***

Move slowly, making sure your upper back and shoulders stay at rest against the car seat. Make the movement as small or large as you like.

Notice that as you arch your back there is a tendency for your chest to rise and for your head to go up toward the roof of the car. Adjust the rearview mirror for your head when it is in the highest position.

4. Keep your back slightly arched in a comfortable place somewhere between arching it fully and pressing it back into the seat. You might find that it takes some work to keep your back slightly arched. Collapsing in the seat causes muscles to weaken, and compresses our spine. A slight arch will give your spine support, which in turn reduces pain and stress. Can you relax your jaw and face while feeling the arch of your back?

✓ ***Awareness Advice:***

The optimal position is where the back feels slightly arched and there is a little bit of work for the muscles. If your back is weak you might find this difficult. But you will find that the muscles develop with practice and the strengthening of these muscles will help you in any seated position.

5. Notice how much easier it is to breathe when you feel the back working a slight amount in a supportive arch with the chest raised a little higher and the head elevated. To notice the difference, collapse your back into the seat, let the chest drop, and let your head go lower. Can you feel that there is not as much room in your chest to breathe?

6. Now sit up again by slowly rolling your pelvis forward to arch your back and elongate your spine. Can you feel how much easier it is to take a full breath? (Many people become sleepy when they drive because they breathe inadequately. That is partly because they collapse their back into the car seat.)

7. You have achieved your optimal driving position when your back has a slight arch, your hands are comfortably on the steering wheel with the elbows down, and you can look in your rearview mirror with a lengthened spine. The key component of your optimal driving position is full, easy breathing. Now, you are on your way!

MOVEMENT LESSON 2

EXPANDING YOUR VISUAL FIELD

This lesson helps to create a soft focus. Most people stare at the road in front of them until their eye muscles become fatigued, their jaw becomes tight, they stop breathing deeply, and they arrive at their destination feeling strained. Remember to learn this lesson on the side of the road until it becomes automatic and you can use it while driving.

1. Find your optimal driving position as in Lesson 1 of this section. Make sure you have adjusted your rearview mirror so you have to lengthen your entire spine to look at it.
2. Look forward and pick a spot in the distance, and let your eyes rest on that spot. Without moving your eyes, can you notice everything to the right of that place? Notice that you can see quite a lot on the right including the interior of the passenger compartment, and perhaps your right arm, without needing to move your eye from the place you have chosen to focus on. Let your eyes become very soft and simply allow the light from the right side to enter your eyes. Close your eyes and rest.
3. Open your eyes and again look at a point in the distance in front of you. Allow yourself to see everything to the left of that point without moving your eyes, including the sky and the lower inside parts of your car. How far to the left can you see without moving your eyes? Remember, don't strain; let the light and the shapes come to you. Rest, close your eyes.

4. Open your eyes again and look to a place in front of you. Notice how much space your eyes can see without moving your pupils from the place you are looking at. Occasionally move your eyes to a new place in front of you and experience the sensation of the environment entering your eyes rather than reaching out with your eyes to the environment. While you allow the environment to come into your eyes notice how that can soften the eyes, the forehead, the face, and jaw.

5. Gradually let your eyes roam, still experiencing the feeling that they don't need to stare or grab at what they are seeing. Take a few slow, deep breaths while you do this.

6. We learn best by contrast so now look at the roadway in front of you and stare at something, squint, and tighten your jaw. Can you feel how hard the eyes become, as well as the rest of your face? Stare even harder and notice that it becomes difficult to breathe fully as the neck and chest become tense.

7. Now soft focus again, let shapes, colors, and light enter you; realize you don't need to reach for it. Feel your jaw as well as your face relax and observe whether it is easier to maintain your optimal driving position with this soft focus that allows everything to be seen at once.

MOVEMENT LESSON 3

EASIER PARKING

Many people strain and even hurt their necks while turning to park. For most people, turning to look behind themselves for a parking space places great demands upon the neck. They stiffen their body to the point where they can turn the joints in their neck. This lesson should help you turn in any situation, not only in your car.

1. Find your optimal driving position, as in Lesson 1 of this section. Then, soft focus your eyes as you look forward, as in Lesson 2 of this section.

2. Look slowly left and right over your shoulders with the intention of noticing which side feels easier. Do not strain, but discover how far you can go easily.

3. Look over your left shoulder. Do you grip the steering wheel harder? Do you hold your breath? Then, look over your right shoulder and observe how the rest of your body responds again. Which side allows easier breathing?

4. While holding the steering wheel, move your shoulders and upper body left and right without gripping the steering wheel harder. Allow your pelvis to participate by turning in the seat as well. (Review Section 1, Lesson 1 if this is difficult.) Pause.

   Continue turning your upper body allowing your pelvis to participate while you keep your face in the center. Look at something in front of you. Soft focus your eyes so that your body turns left and right while your head remains stable. Rest.

5. Continue turning your body as in Movement 4, but this time, take your head slowly in the opposite direction at the same time. If your body turns left, your head turns right. If your body turns right, your head turns left. Rest. Return to simply turning your head to look behind yourself. Now, see if it is easier when more of yourself can participate in the turn.

6. This time, find a place in front of you to soft focus. Remain looking at that point and slowly turn your head left and right. Do this until it is easy to breathe while doing it. Pause.

   Can you intentionally move your eyes to the left while your body turns right and vice versa? Make sure your body is relaxed and your face is soft. Pause. Simply turn your face left and right taking your eyes with you. Do you have more movement in your neck?

7. While in your optimal driving position, simply turn left and right again, as if parking your car, and notice if it is easier when you have so many more parts to move. Notice how much of yourself you can mobilize to turn.

SECTION IV

Dealing with Back Pain

Before you begin this section, make sure that you are as comfortably clothed as possible. Remove anything that is constricting. Take off your shoes, take off your belt. If you have any jewelry that can catch in the carpeting, take that off also. If you are warm enough, you can even remove your socks. The less you wear, the better. Choose a quiet piece of the floor that has a carpet or a mat. Make sure you have enough room around you so that if you lie down on your back and stretch out your arms and legs, you will not feel constricted.

Most conventional back programs contain back exercises that require moving in a very linear manner. For example, always bending straight forward or backward or doing exercises to stabilize and protect the vertebrae of the spine, which requires stiffening the entire torso. The exercises in this section combine several directions of movement at once, which is how we normally move. For example, when someone bends down to pick up a napkin, or bends over while sitting in a chair, he or she actually combines movements that are slightly forward and partly to the side, with rotation, all at once. When we reach up to paint a ceiling or put something on a high shelf in our kitchen, we must include complex movements of our pelvis that balance us on our feet with the action of lifting up our arms and looking at what we are doing. Many muscles and joints are involved in seemingly simple actions. Increasing the flexibility of only the shoulders or neck still puts too much stress on parts of the body that are not involved in the motion in an obvious way. To include all body parts with every action, the lessons in this section will link the head, chest, and pelvis together. They require being linked as we go about our day.

MOVEMENT LESSON 1

BODY SCAN

This lesson is an excellent way to reduce tension without any movement, ideal for those times when you are too tired to move. Some people use this lesson as a means of focusing the mind on the body so as to rejuvenate themselves as they might in a meditation. Others use the lesson before every lesson in this book as a way of tuning in to themselves before performing any particular lesson. Eventually, you can go through this lesson in a few minutes. If you do this on a regular basis it could develop into a good habit.

1. Lie down on your back. Put your arms down at your sides, and lie with your legs long, and your knees straight. You might find that this position hurts your back. If it is uncomfortable, bend your legs so that you can put the bottoms of your feet flat on the floor.

2. Your first exercise is to observe how your body contacts the floor. Use the floor as a kinesthetic mirror to teach you how your body feels. Observe the way that you're resting on the ground right now, the posture that you have on the floor. For example, is there some part of your body that's pushed heavily into the ground? Is there some part of your body that seems arched, high off the floor? Notice how the floor supports you. Are all of your muscles relaxed enough to allow as much support as possible from the floor?

✓ ***Awareness Advice:***

Many people confuse position with posture. Your posture is an ongoing dynamic process that expresses itself in the way you stand, sit, or lie on the floor. In other words, posture expresses itself in every position of the body. This is why people tend to have tension in the same muscle groups regardless of the position they are in. If you observe your posture on the floor, you will notice basic postural tendencies that are expressed in any position.

3. Feel the pressure under the back of your head and notice the shape this pressure makes on the floor. Also notice the shape of the pressure on the back of the ribcage and on the back of the pelvis. Which has the largest area covering the floor: the head, ribs, or pelvis? Is there more of yourself covering the floor on your right or left side?

4. Notice the curves of the spine, in the lower back and the lumbar spine, in the neck, and in the cervical spine, where you do not have contact with the floor. Which one of these two curves seems higher from the ground? Don't put your hands there to tell. Try to keep the hands down at your sides and use the pressure from the floor and your own kinesthetic sensations to guide you. Which arch is higher, the neck or the lower back? And which has a longer span, the curve across the neck or the curve across the lower back?

5. Observe the way your legs rest on the floor; you might discover that very little contact exists between the pelvis and the calf muscles. Many people will hold their legs off the floor so the hamstrings don't touch the ground. This is because of unintentional and unfelt work in the hamstrings. Next, notice the shape of the pressure under the calves and under the heels. One of your feet and legs might feel as though it's turned open more than the other.

6. Notice the relative height of the shoulders. You might discover that one shoulder is higher from the floor than the other, one shoulder blade might feel broader, more spread, than the other. If you look at the back of the ribcage and feel its pressure on the floor, you might notice that it's rolled or tilted more toward one side than the other along with the shoulder. Very few people are symmetrical here.
7. Imagine that you are lying with your spine on a balance beam or tightrope. Which way does it feel as if your body would fall off? Observe separately which way your pelvis feels as if it would roll, and then your chest, and then your head. Don't be surprised if all of yourself doesn't go one way.
8. Notice how much of the back surface of your body is actually on the ground right now. If you had to take a guesstimate, and you looked at 100 percent of the back surface of yourself, what percentage is actually touching the floor?

✓ ***Awareness Advice:***

Some people might find 20 percent, others, 80 percent. It could vary that much, depending on how much muscle tone you have in the back of your body. The tighter the back muscles are, the higher you'll be arched off the ground, and the more you'll be lifted away from the floor. If the extensors in the back are tight, it's like tightening a bowstring, and, of course, the bow curves more. One thing I'd like you to notice as we do these lessons is whether you begin to lower, flatten, and spread giving more of your weight to the floor.

9. Now let's discover why lying with long legs hurts some people's backs. Bend your knees and put your feet flat on the floor, and notice what has happened to the position of your lower back. Does it feel closer to the ground? Is the pressure in a different place under your pelvis? Now slide the legs long again and notice the changes in the back and in the pressure under your pelvis, and then bend the knees

again to rest flat on the feet. Do that a couple of times until you can go back and forth and know clearly how the position of the legs affects the position of the lower back and the pelvis.

✓ ***Awareness Advice:***

As you do the lessons in this book, but particularly the lessons that are done lying on the floor, you will experience changes in the configuration of your body against the floor. Some parts will feel longer, looser, and flatter against the floor allowing more support from the ground. Frequently notice the changes whenever you are resting or pausing between movements in any of the lessons.

MOVEMENT LESSON 2

PELVIC ROCK AND ROLL

This is one of the most familiar movements performed in back programs. However, it needs to be done with full awareness in order to maximize the benefits. Remember, these lessons do not just provide movement that is good for your body but provide information that is useful to your brain.

1. Bend your knees and put your feet flat on the floor. Put your feet in a very comfortable place, so that they are about shoulder-width apart. Make sure that the ankles are not near each other. With the feet shoulder-width apart see that the feet are flat enough on the floor that you can feel your heels and your toes on the ground. Have the knees balanced above the feet.

✓ ***Awareness Advice:***

You might find that there is such an arch in your neck right now that your head is hanging back and it's hard to breathe, and you feel uncomfortable. Get a cushion, a pillow, or a towel and put it under your head so that it is raised to a comfortable enough height to relieve the stress in your neck.

2. Push your feet into the floor so that your pelvis rolls toward your head and your lower back goes closer to the ground. If it feels like it's completely on the ground, then think of simply pressing the lumbar spine closer to the floor. Now

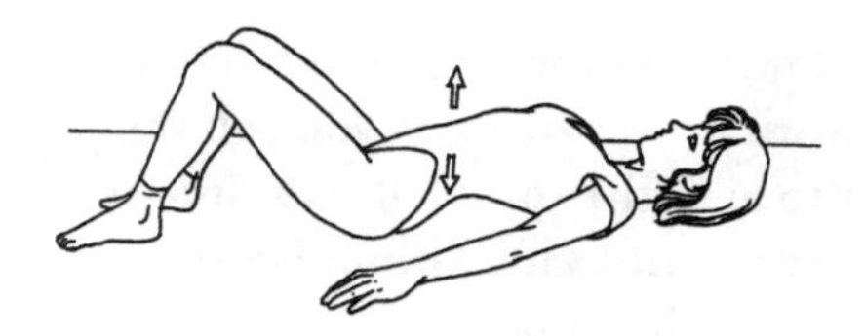

release that and roll the pelvis toward your feet; you'll feel your weight moving down toward the bottom of the pelvis as your back contracts. When you do this, the lumbar spine will arch up off the floor. Push again with your feet and roll the pelvis back toward your head until the lumbar spine goes flatter. Roll your pelvis back and forth from head to foot, while noticing how your lower back flattens and then arches. Take a rest.

3. Do this same movement resting your hands on your belly. With the hands on your belly you might notice that you're contracting your stomach to help lower your back to the floor. Try not to do that. Learn how simply pushing through your legs into your feet is enough to help direct the pelvis to roll toward your head. Make sure you're not lifting your pelvis off the floor. As it stays heavily on the floor it will roll far enough that you feel the lower back go flatter. Then, roll the pelvis toward the feet while feeling its weight heavy on the ground. You'll notice a curve as the lumbar spine arches. Go back and forth fully within the limits of comfort. Take a rest.

✓ ***Awareness Advice:***

While doing the movement, you'll discover that your head is pushed and pulled as the spine rolls. It pushes the spine toward the head and then pulls the spine down away from the head. You should notice that your head on the floor is making an up-and-down motion. Release the jaw, let your teeth come apart, soften your eyes, and see if you can roll your pelvis further in both directions. Go very slowly. Make sure to go slowly enough that you can feel your legs

pushing down into the feet, into the floor. You can feel the work in your back as you roll the pelvis toward your feet to arch the spine. If you find one of the directions uncomfortable, diminish the range of motion in that direction.

4. Return to the previous movement, but see that it can be done by alternating the work between your legs and your back, so the stomach feels independent. It doesn't need to contract. Often we learn to clench our guts and stiffen our belly whenever we want to move our back in a certain way. Begin to take a slow, deep breath in your belly, and as you do that, speed up the motion of your pelvis. See if, in fact, you can make several movements of your pelvis, back and forth, for each breath. If you can take a slow belly breath and make at least five pelvic movements in both directions, easily and comfortably, then your belly is probably free from the tension of the motion. Now, pause and take a rest.

5. Take your arms and rest them on the floor above your head, so the backs of the hands are resting against the floor or as close to the floor as you can get. Feel the chest expanding and the ribs being more open. Once again, rock your pelvis back and forth, from head to foot, and notice if it feels like a larger movement to you now. Can you feel that the position of the spine changes? Feel if you observe more range of motion for the pelvis now.

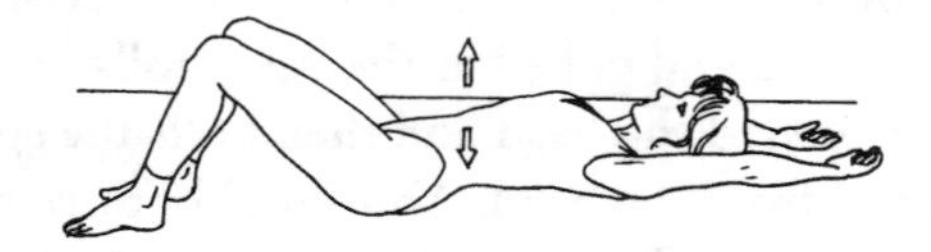

✓ ***Awareness Advice:***

Remember to make the movement as large as you can comfortably make it. Go as fast as you can comfortably go, once you're certain of the movement. If you've gone slowly, you're very certain of the movement. Begin to go faster, like a train taking off from the station. It gradually accelerates. See if you can do that without any tension in the belly, so that no matter how fast you're going right now, your breathing stays very slow. That's important. Pay attention to that. Make the breathing long, deep, and slow and make rapid movements of the pelvis back and forth. Feel the head being free to respond to the movements of the pelvis, let the jaw relax, let the head nod up and down as it rolls on the floor, and feel the chest being soft. Make sure you feel your feet pushing to help roll your pelvis to the head, even if you're going quickly. And leave it. Take a full rest. Let your legs go long; let your arms come down.

6. Once again, with your legs long give your back to the floor. Notice if the contact that your body is making with the floor has changed from the beginning. See if you feel lower now, flatter, more spread, and more giving to the ground. That simple pelvic rock may have been effective enough to lower your muscle tone quite a lot. Some people find that common, ordinary back pain can be quickly alleviated with the pelvic rock.

MOVEMENT LESSON 3

ELBOWS TO KNEES

This lesson will provide an interesting way of folding or flexing your body. During the lesson, you will feel many changes occur in the muscles of your back. As with any learning activity, the less you strain while performing this lesson, the faster you will learn and the more you will retain. Inversely, the harder you strain, the more this becomes just an exercise and the less it becomes a lesson whereby you can learn to control the muscles of your back.

1. Gently lift your head to look toward your feet. Notice the amount of effort it takes to perform this movement. Don't strain. Let the head down again. Do it another time or two, only to notice how much effort is involved and how high the head can go without struggling. How high does it go easily? Where is the work taking place? Now let the head back down. Remember how high it went easily and remember the feeling of the effort because that will change as you improve the use of your belly and chest and allow your back to release resistance. Take a full rest.

2. Bend your knees so that both of your feet are flat on the floor again. Take your right hand and put it underneath your head. Feel the weight of your head in your hand. Now, draw your right knee above your chest so the foot comes off the floor, and hold onto the right knee with your left hand. If you were holding directly onto the knee, you'd be holding the part of the knee on which you feel pressure when kneeling. See that you have as much

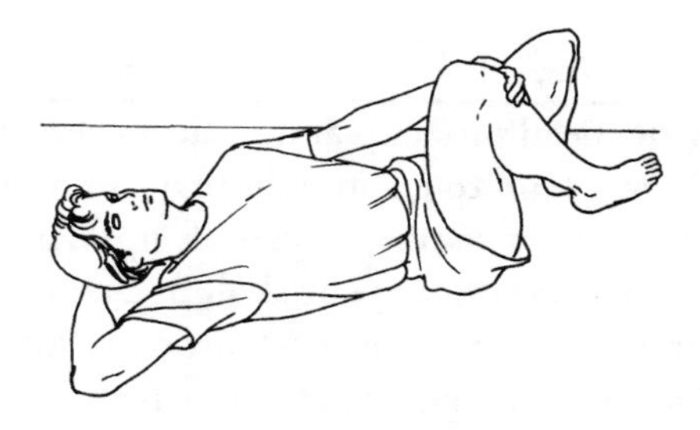

distance as you can comfortably maintain between the right elbow and the right knee so you are completely relaxed. Therefore, the right elbow should be resting on the floor and the right knee should be far enough away from you to keep the left elbow straight.

✓ ***Awareness Advice:***

If you cannot easily hold on to your knee, put the hand in the crease in back of your bent knee, or even hold on to some part of your pants.

3. Now very gently and slowly, draw your right elbow and your right knee toward each other, picking up the head as you do so. Go toward the knee without any attempt to touch it, and feel how the ribs will tend to close like accordion pleats. This is a time for the air to squeeze out of your body. When you set the head and elbow back to the floor, the ribcage expands and it's easier to inhale. Move at the speed of slow, steady breathing.

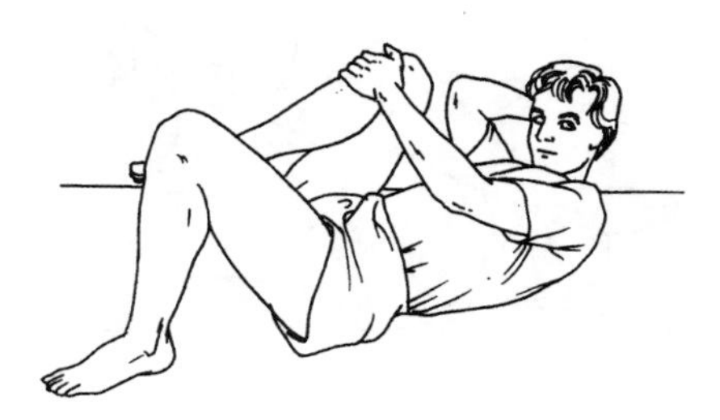

✓ *Awareness Advice:*

Feel how the head picks up and notice where it wants to go. Does it tend to go straight forward or off to the side? Does it want to turn as you lift up? Are you putting your hand under your skull in a comfortable place that will support your neck properly for you? You will have to discover what works best.

Notice the distance between the elbow and the knee. Your knee and elbow may be touching—that's fine as long as you are comfortable and there is no effort. Feel the part of your belly that is working to draw the elbow and the knee together. Now make sure that your elbow is pointing directly to your knee instead of anywhere in the room.

Repeat drawing the right elbow and knee together several times. Then rest with your arms and legs long and observe any differences between the two sides of your body, especially your back, shoulders, and hips.

4. Next, bend your knees again and place your feet flat on the floor. Put your right hand behind your head, as before, but this time, raise your left knee up and hold on to it with your left hand. Draw the right elbow and the left knee toward each other. You'll discover a different use of the muscles in the chest and belly than in the previous movement.

✓ ***Awareness Advice:***

Go slowly, easily. What does your head want to do? Are you stiffening your neck and insisting that your head point toward the knee? Or do you notice that as the right elbow moves toward the left knee, the head wants to turn, all by itself, further to the left? Can you allow it to turn? Make sure that you always increase the distance between the elbow and knee by setting the right elbow back toward the floor. Never let go of the left knee.

Rest on your back and again observe differences in your contact with the floor.

5. Now see if you can make the other side of your body feel just as good. Begin by bending your knees with the feet flat on the floor and the left hand under your head. Raise your left knee and hold on to it with your right hand. Again, make sure that you are as comfortable as possible. Then, draw your left elbow and left knee toward each other several times and explore if this side is different in effort and flexibility then the other side. Rest and feel the change in your back, shoulders, and hips.

✓ ***Awareness Advice:***

The harder we work, the less we can feel differences, because we tend to muscle our way through difficult places. Awareness requires less effort so you can feel more clearly what distinctions and differences there are in your movements. Also, improvement and clarity occur in your movements if you're more precise about what you are actually doing. So, each time you take your elbow and knee toward each other, imagine an ant crawling toward two different places on your knee so that your elbow can direct itself toward a small and precise point that varies as the imaginary ant moves. This variability in the directions of your movement will cause new muscle fibers to be engaged with every contraction.

As a way of demonstrating the effectiveness of variability, bring your underarm and knee together a few times, and then return to taking the elbow and knee together. You will probably find an immediate increase in the range and ease of the movement.

✓ ***Awareness Advice:***

How fast should you do this? As fast as the speed of your breathing.

6. Again, hold the right knee with your right hand and put the left hand under your head and draw them together. Search for comfort, ease, and the path of least resistance. Rest fully with your arms and legs long.

7. Interlock your fingers and place them underneath your head as if you were resting your head in a big basket. Bend both of your knees and draw them above yourself. Make sure your feet and ankles are relaxed, and that your pelvis remains on the floor. Your knees should be somewhere above your abdomen. Very slowly, close both of your elbows toward the sides of your head, then open them on the floor again. They will be opening and closing like a butterfly's wings.

8. The next time the elbows close toward your head, draw both elbows and both knees toward each other, and then let them go completely apart. As the head and arms rest back on the floor, the knees will go a little bit away from you again, and if it's comfortable enough, you could let your feet return to standing on the floor. Draw both elbows and both knees together and apart several times, learning to breathe while doing so. Rest on your back with your arms and legs long.

✓ ***Awareness Advice:***

If you find your neck is straining from pulling on the head too hard, move your hands further down toward your neck, so the neck itself is supported throughout the movement.

9. Now draw both elbows and knees together again to the fullest comfortable distance. If you're touching, remain with the elbows touching the knees. If you're very flexible, you might be able to grasp your knees with your elbows. Otherwise (and for most people) find a comfortable relationship and then imagine that you are supporting a stick between each elbow and knee. Without changing the configuration that you're in, very slowly roll a small distance from side to side. Go as long as you want, but make sure you can still breathe.

Rest fully on your back with your arms and legs long and observe your changed relationship to the floor. Working the muscles in the front of our bodies causes the muscles on the opposite side, in back, to release without forcible stretching. When you feel ready, walk around slowly and observe if anything in your posture has changed.

MOVEMENT LESSON 4

Comfortable Twisting

This lesson will give you an opportunity to rotate your spine and free the muscles on the inside of your legs. When we suffer from pain and stress, we lose the ability to have independent motion of our pelvis and ribs, and we tighten the inside of our legs by pulling them together protectively. This lesson is a pleasant antidote.

1. Lie on your back and take a moment to feel your contact with the floor. You might want to review the body scans from the beginning of this section.

2. Put your arms away from you, out at your sides, so that the hands and the elbows are the height of your shoulders. You might want to roll your head and look at your hands, making sure they're at the same height as your shoulders. Just rest the arms on the floor. Bend your knees with your feet flat on the floor and shoulder width apart. Now, open your knees away from each other so that you roll toward the outside edges of your feet a bit, and then pull them together a bit. You'll feel the adductors being stretched as you open the knees, and they'll contract as you close them (Note: the adductors are the muscles on the

inside of your legs). In this rotational movement, we're going to work toward opening the adductors more.

3. Keep the knees together and roll over your feet so that both of your knees point toward the right. Let them hang to the right. Now, take your left knee and pull it to the left, keeping the foot on the floor. Roll over the foot and pull it away from the right knee as far as you can go, making sure that you do not move the right knee, and then bring the left knee back to the right. Go back and forth several times with the left knee, making sure that the right knee never moves. You might even want to touch your right knee to make sure that it's not moving at all, and see that each time the left knee goes, it can go a little bit further before the right knee moves. We have a deep habit in our culture of clenching our adductors together, which misorients the spine and compresses the pelvis. So see if you can find a way to teach yourself that the spread between your legs, the opening, can increase more than you think. Now, take a rest with your legs long.

4. The next time you pull the left leg to the left, keep on going until the right knee has to follow and then point both of your knees to the left side. Now take your right leg and pull it away from the left leg as far as you can. See if you can overcome those deep habits and allow the legs to open much farther than they're used to. Then go back and forth with the right leg a couple of times, making sure the left leg stays still. Rest.

5. Now alternately take your knees from side to side with one leg pulling actively until the other leg must follow. Keep the feet heavy on the floor; notice how the pelvis rolls as the knees go from side to side, pointing to the left and

then to the right. While your knees go from side to side this way, with one leg actively pulling the other leg, allow your head to roll in the opposite direction so that you can feel a pleasant twist through the spine.

6. See if you can let your eyes look at your hands. Therefore, as both knees are at the right, try to look at something on your left hand—you may have to stick a finger up to do that. As both knees point to the other direction, your head will roll to look at the other hand. During the movement, the knees should be close together while at the sides and wide open when they're facing the ceiling. Let it be a sensuous movement. One knee actively pulls, the entire body opens to the ceiling, the entire pelvis is open, and the face is to the ceiling. Then twist to a close at the sides, with the knees together, and the head in the opposite direction. Open to the ceiling once again, and then close in a twist to the sides. The farther you rotate the body to the sides, the more you'll feel the ribs becoming involved in a pleasant stretch. Feel the opposition and the coordination.

Now leave it. Let your arms and legs lie long. Take a full rest. Take a minute here to enjoy the pleasantness of your body's contact against the floor, the fullness of it. When you're ready, stand up and walk, observing the change in your body.

MOVEMENT LESSON 5

EASY BENDING

Many people find that they cannot bend backwards very easily, that they lack both flexibility and strength. This lesson will provide you with both, plus give you the awareness you need about how your lower back and hips are connected to help you bend backwards.

1. Rest on your back with your arms and legs long. Feel your contact with the floor and observe which parts of your spine feel as though they could bend backwards the easiest. Can you imagine this while you are lying on your back?

Now roll over and lie on your stomach; let your legs be a comfortable distance apart. Place both of your hands on the floor near your head with your elbows wide apart and your fingertips almost touching. Place your chin or forehead on the ground and imagine that you can see an imaginary insect on the floor just above your hands. Raise your head so that you can see it. Now watch it run away from you, up the wall in front of you. There it goes. Now watch it run up to the ceiling. Keep watching it run on the ceiling as far as you can. Let your chest peel off the floor and then use your arms to help you, so that you can look up a little further. Do not move your hands, no matter what. Practice this two or three times.

✓ ***Awareness Advice:***

As you are watching this imaginary insect, you'll begin to feel the back of your head pulling backward toward your pelvis. Notice the front of your body beginning to peel off the ground: the chest, the lower ribs, and the belly. You can use your arms to help you by pushing with them. But make sure that you use the back as the primary initiator of the movement. Go up and down several times. If you watch this imaginary bug, you'll discover that your head starts the movement and then the chest will begin to peel off the floor before you engage your arms. So the back is used first. Take a full rest.

2. Put your hands into push-up position so the hands are wide away from the shoulders, elbows up toward the ceiling. Watch the imaginary bug again. This time, as you look up, feel that the head comes up first. Don't just push with your arms. The arms are there to assist and support the spine, but it's the back itself that begins the work and does most of it. Let your arms straighten if you can, but notice a feeling of gradually peeling the front side of your body off the floor as you look up. Go up and down this way, peeling yourself off and sticking yourself back on to the floor. Rest.

3. Repeat the previous movement again, only this time let your head hang completely down. See if you can lift up your body with your head dangling. Try it as though you were looking at your chest and your belly. After doing this a couple of times, let the head lead the movement up again.

Now, rest on your stomach and turn your head and arms in any way that's comfortable. If you feel some stress in the back, toss your pelvis from side to side freely, letting the heels flop. Rest.

4. Now turn your face to your right. Lift your right leg two inches off the floor with a straight knee, and then set it down again slowly. Lift your thigh and your knee only two inches off the floor. Can you feel your shoulders being affected as your leg moves up and down? As you lift your leg, can you feel your chest being affected? Be sure to move easily and slowly so you can feel how lifting your leg affects your back. Rest.

5. Turn your head to the other side, facing left. Can you imagine your left leg lifting only two inches off the floor? Can you guess how that would feel in your upper back and shoulders? Now, lift your thigh two inches off the floor. Remember it's only a small amount. Is this leg easier to lift than the other, or harder? Rest.

6. Lift the leg again a small amount off the floor. Feel the leg reaching back for an imaginary canvas. Now, with your toes, paint tiny circles in the air: not with your ankles, but with your hip so you feel that your entire leg is the handle of a paintbrush and your toes are the bristles of the brush. Feel how round you can make your circle. Rest.

7. Turn your head to the other side. Now imagine the left leg doing the same movement. Lift the thigh two inches off the floor and reach your toes and foot toward this imaginary canvas, to paint circles. You'll feel the buttocks working a lot. Be sure to keep the knee straight.

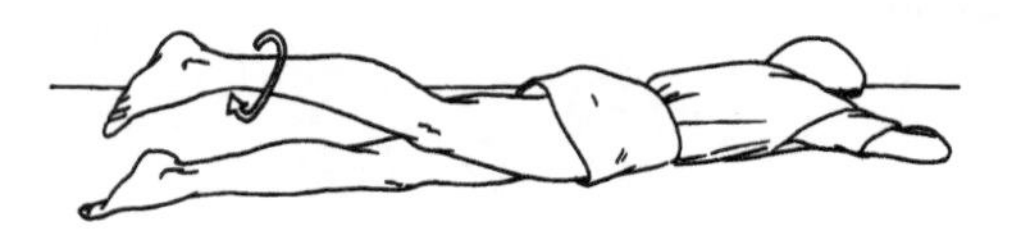

✓ ***Awareness Advice:***

The rounder the circle, the more you'll awaken all the muscle fibers in the gluteal area.

Rest when you need to and go in the other direction with your circle. You might find that one direction is harder than the other. Explore this on the other leg as well. Leave it and take a full rest.

8. Now, put your hands under your head so you can rest your chin on the back of your hands with your palms on the floor. Rest your chin, or even your mouth, in any way that's comfortable on your hands. Now, think carefully before you move. Lift your head and both arms together off the floor, the elbows, the hands, and the head. That means the arms and the head are together, as though glued. Go slowly; feel how your back works. What you'll feel here is, again, a peeling of your chest off the floor. As your hands are together, you should feel that your elbows begin to pull backward, up toward the ceiling, to help you go higher. Is it easier if you let your legs lift a little bit? Rest.

9. Can you lift your arms and head again as one unit, staying up in the air? Can you leave your head up in the air and lower your arms back down to the floor, and then raise them back up again to greet your chin? Raise and lower a few times. Can you feel the effect on your legs? Rest completely and toss your pelvis from side to side in order to release any strain.

10. Finally, look up to the ceiling again, watching your imaginary insect as you did in the beginning of this lesson. Is it easier?

Roll over and rest on your back. Now, while on your back, you might discover that you're arched more off the floor than you were. Notice where your spine happens to be; it's not important if it's arched right now.

MOVEMENT LESSON 6

REGAINING FULL USE OF THE NECK, I

1. Lie on your back. Put the palm of your right hand on your forehead with your elbow pointing out to the side. Imagine that you have a board from your fingertips to your elbow and that the head is a cylinder or a ball. Now, roll your head underneath your hand by pushing with your arm and hand. As you roll the head to the left, you should be pushing down toward the wrist end of your hand, and as you pull the head back in the other direction, you should roll toward your fingertips.

✓ ***Awareness Advice:***

See that the head feels completely passive, that there's no activity or work in the neck, and that it's all being done by the power of the arm. You should feel the skin on your forehead sliding across the bone of the skull before you feel any movement in the head. If you feel the skin pulled to the limit across the bone, and finally the head beginning to move, you'll know that the arm is doing the work. Let the jaw relax, the eyes go soft. Now, set the arm down. Leave it and rest.

2. Put your left hand on your forehead in the same manner as before. Move your head with your left hand in the same way as you did with your right. Make sure the left elbow points to the left. See that you minimize work in the neck.

✓ *Awareness Advice:*

Ideally, you want no activity in the neck at all. The head and neck should move passively, as though someone else's hand and arm is doing this to you. Some people find that this is an excellent exercise when their neck is so tense or so painful that they actually can't turn it. If they can learn not to compress the joints by contracting the muscles, they can gain mobility and freedom.

Leave the movement. Let the arm rest at your side. See if the quality or condition of the neck is different.

3. Now, roll over onto your stomach and put your forehead on the floor with your hands in push-up position. This means that the hands will be a little wider than your shoulders and the elbows will be pointing toward the ceiling, as though you were going to push up. See that your feet are apart.

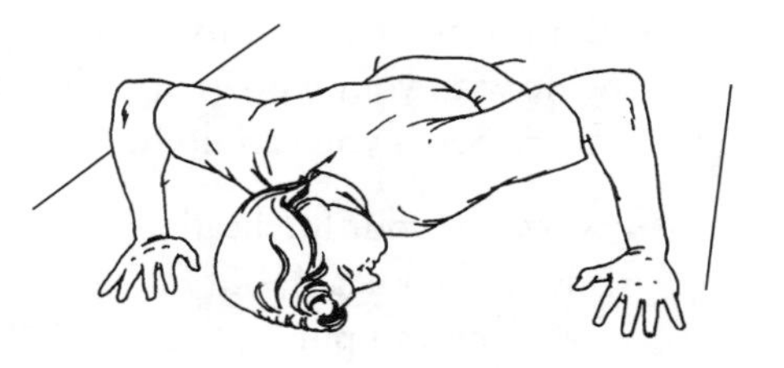

4. Find a way to roll onto your left cheek and across your forehead by pushing through your right arm. Then, roll across your forehead to the right cheek by pushing through your left arm. Take your time and find a way to roll gently and slowly across the forehead from cheek to cheek, so that you can look out at your sides. Now leave it and rest your head and arms in any comfortable position. Stay on your stomach.

✓ *Awareness Advice:*

As you do this, feel that the arms provide support for your neck. Imagine that your head is a ball rolling at the end of your spine. As you push your ribcage and spine from side to side with your arms, your head will naturally have to follow. See that your pelvis stays flat on the floor. It's the spine and the upper body that are moving from side to side. If you're rolling across your nose, simply lengthen the back of your neck and you'll discover that you don't need to. The longer you make the back of your neck, so the forehead can be on the floor, the less pressure will be on your nose.

5. Put the arms back into push-up position. Have your hands wide with your elbows up. This time, put your chin on the floor. See if you can do it.

6. Now, put your forehead on the floor again, but see if you can put yourself higher up on your forehead, more toward the top of the head than you were before. Notice what happens in your chest. Now, can you put your chin on the floor, setting it farther away from you than before? Alternate between the forehead and the chin, very slowly. As you do, see that you move farther up on your forehead each time, as though you wanted to look under yourself, toward your belly. When you set your chin on the floor, set it farther away from your body, so you can look out farther.

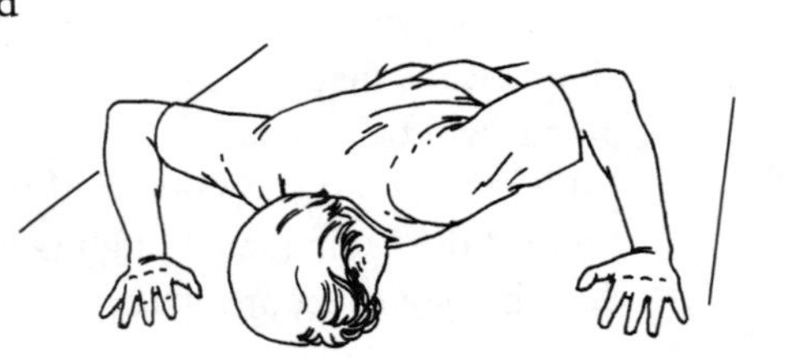

✓ ***Awareness Advice:***

As you go back and forth, notice the movement of the chest and the lower back. If you reach out a long way with your chin, you'll find that there's a pulling that comes all the way through the pelvis. Notice how the legs are involved in the movement.

This particular movement is very useful to people who can't bend their heads at all. If they find they can't flex or hyperextend their neck without pain, they often can in this particular position. Gravity is providing a certain support to the back of the neck, and more motion can be obtained lying on the stomach.

Now leave it and once again rest on your stomach in any comfortable position you can find. If you can't rest on your stomach, then of course lie on your side or on your back.

MOVEMENT LESSON 7

Regaining Full Use of the Neck, II

Make sure you have done the preceding lesson before you attempt this one. Some people think of this as a posture-control lesson because it counteracts the tendency to slump by making you aware of the relationship of the shoulder blades to the ribs and head.

1. Lying on your stomach, interlock your fingers and put your forehead on the floor. Put your interlocked fingers on the back of your head, so you feel the weight of your hands on the back of your skull with your nose on the floor. There should be no pressure on your nose; all you need to do is lengthen the back of your neck in order to eliminate this pressure. If you have a shortened back of the neck, arch it and you'll find you'll press on your nose. If you lengthen it, you'll decrease that pressure.
2. With your hands on the back of your head, lift your elbows and then set them down. Do this several times, up and down, and feel which muscles you use. You'll notice that the back of the shoulders and the muscles between the shoulder blades are used. This area tends to be quite weak in people. This weakness can cause the shoulders to fall forward with the upper part of the body slumping; you may have noticed yourself doing this when you're very tired. This movement will strengthen this area.
3. The next time the elbows are up and you feel that they're level, keep them there. Imagine that from elbow to elbow,

through both arms and the interlocked hands, you are a board, and the board is set on top of the back of your head. Now, push into your head with your right hand and pull your left elbow to your left, so that your head rolls onto your left cheek. Then push the other hand into the side of the head and go to the other side.

4. Begin to roll across your forehead from cheek to cheek, keeping your elbows as much off of the floor as you can possibly keep them. See if you can do it. How can you keep the elbows off the ground? Roll from cheek to cheek across your forehead, and as you do this, feel that your arms move from side to side while your head rolls almost passively underneath them.

✓ ***Awareness Advice:***

Decrease the stress and work in the neck. You should feel that, as you roll from one cheek across the forehead to the other cheek, your shoulders and arms are moving to the side. The ribcage expands and opens on one side and then closes on the other. The more you put your attention on the work in the upper back, ribs, shoulders, and arms, the less work you'll feel in the neck. One thing you might do is see if you can look underneath one of your arms out to the side. Keep going that way, always trying to look underneath your elbows. You'll discover a lot of work in the muscles of the lower back and you'll feel your pelvis working to stabilize and counterbalance the upper body's movements. Feel free to take a break from the movement if necessary.

MOVEMENT LESSON 8

Better Use of Back and Hips

This lesson uses movements that are similar to some of a baby's first movements as it's learning to coordinate its hip and leg motions with its back. Many people find a considerable release in the tone of their hips and belly as well as in their back.

1. Roll onto your stomach. Rest the palms of your hands on the floor above your head, but near it, and near each other so that the arms form a triangle. Turn your face to the left so that you're resting on your right cheek. Now toss your pelvis from side to side freely. Feel how your pelvis rolls from hip to hip. Let your heels flop from side to side also. You should notice them wanting to turn from side to side to help you. Take a full rest.

2. Very slowly lift the left side of your pelvis off the floor so you roll toward your right hip, and then slowly set it down. Repeat this several times and as you do, make sure you're using your back to lift that side of your pelvis. Make sure you aren't pushing into the floor with your foot or your knees. In fact, if you're using only your back, you'll hardly be aware of what you're doing at all for awhile.

3. You should feel that the left leg becomes softer. The higher you lift the pelvis on that side, the more the left knee will tend to bend. Allow that to happen. Also, as you lift the left side of your pelvis and roll your weight to the right a little bit, you'll find that the left knee will turn toward the

inside. Allow it to do that. The higher the pelvis goes, the more the knee softens and bends and rolls toward the inside. Repeat this movement many times. Pause and take a rest.

4. The next time you do this movement, take advantage of the fact that your left knee bends as you lift the pelvis anyway. Pull your knee up toward your waist. Lift your left hip and slide the inside of your knee on the floor up toward your waist, so the leg is bent. This is a position many people find comfortable to sleep in. Now, straighten the knee by pushing your foot straight back down the same way it came up, as the hip lowers. Take a rest.

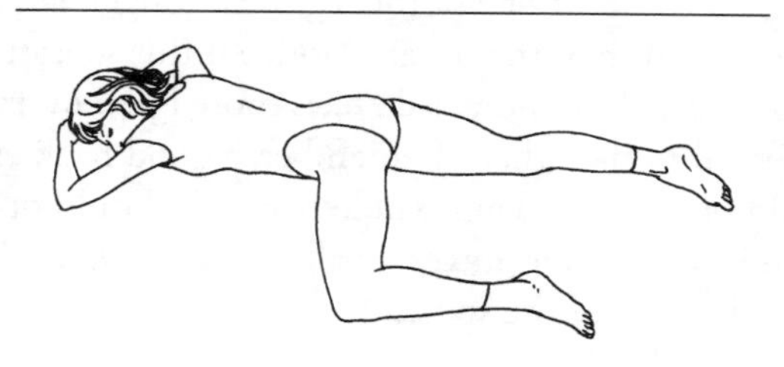

5. Take your left hand and put it into a push-up position, next to and away from your shoulder, so that if you needed to you could push with it. Now, once again, lift the left side of your hip, drawing the inside of your left knee on the floor up toward your waist. Keep the foot on the floor as you do this, and then slide the foot straight back down along the line you pulled it up. Now, slowly go up and down that way several times.

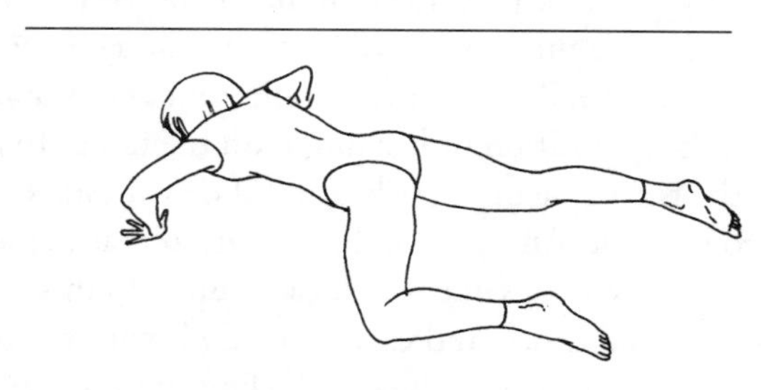

✓ ***Awareness Advice:***

What you might notice here is that if you allow the back and the pelvis to roll more, and if you allow the back muscles to work to help lift the pelvis, it becomes easier and easier to slide the knee. See what you can do to reduce friction here, to lighten the load on the leg, so you don't feel as much pressure on the knee.

6. Roll over and take a full rest on your back. Feel the difference between the two sides of the body. Notice the contrast between the two legs and between the two sides of your pelvis. See if one of the legs feels longer, larger, or more relaxed than the other. Notice if this sensation is also true in the pelvis: if one side feels more open or expanded compared to the other. Remember, a good back exercise should be good for your whole body. It should open, expand, release, and integrate all of the body, not just some particular part of the spine.

7. Roll over onto your stomach again. Let the legs be a comfortable distance apart. Put your hands on the floor near your head, and turn your face to the right. This time, lift the right side of your pelvis off the floor so you roll your weight to the left. Don't pull the knee up yet; simply learn how to lift the right side of your pelvis and set it down. Discover how your back assists the sliding up of your leg. Now pause.

8. Lying on your stomach, put your right hand in push-up position. If you found that this movement was very easy on the other side, you might want to leave the hand on the floor above your head. You have a choice of either standing on it in push-up position or keeping it on the floor. In any case, lift the right side of the pelvis and slide your knee up toward your waist. And then slide it down so that the foot goes straight down the same way it came up. Do this movement several times, and each time, do it with the foot going up and down that track, and the knee pulling up and going down. Discover how you can make it easier.

✓ ***Awareness Advice:***

How can you make the movement lighter? How can you coordinate the back, the hip flexors, and the muscles deep inside the stomach so they all work together to make it a simple and light movement? See that you don't stiffen the neck, and that you don't hold your breath. Let the jaw relax.

Leave it. Roll over and rest on your back. Enjoy the openness of your hips and the softness in the legs and belly.

MOVEMENT LESSON 9

Leg Tilt

In this lesson, you will learn to have more independent motion of your spine. Many people cannot turn their chest and pelvis independently, especially if they have pain or only practice doing exercises that build strength in particular muscle groups without learning to develop independent motion of each of their vertebrae. This lesson will give you an increased awareness of your waistline and an integration of the upper and lower halves of your body.

One caution: if you have had hip-joint replacement surgery, do not cross one leg over the other while doing this movement. Only do the first five parts of the movement in this lesson.

1. With your arms and legs long at your side, feel your contact with the floor. Take a guess: of the entire surface of the back of your body, head to heels, what percentage is touching the floor right now?

   Now, imagine you're lying on a balance beam that is along your spine, dividing you into two halves. If you started to fall, to which side do you feel you would roll off the beam? You might find that your pelvis would roll in one direction while your head and chest would roll in the other. If there were an emergency, and you had to roll out of the way as fast as possible, to which side do you imagine you would roll automatically?

2. Bend your right knee so the right foot can rest solidly on the floor, not too close to yourself and not too far.

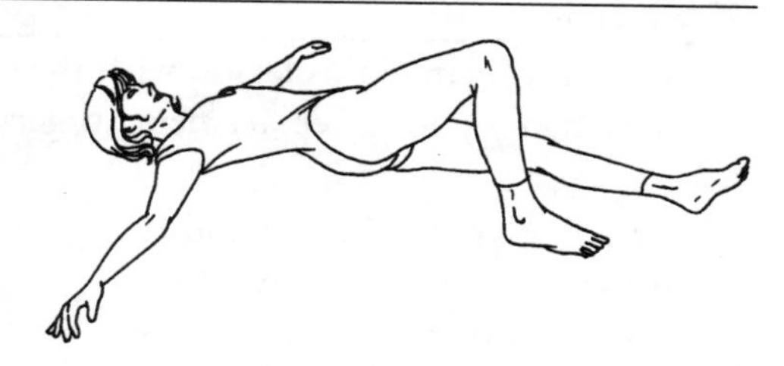

Find the balanced place where it takes the least effort to stay in this position. Push down through the right foot so the right hip lifts off the floor a little and you roll to the left. Let the movement be very small and make sure that the left leg is relaxed. The right knee should stay directly above the right foot during the entire movement of the hip. Experiment with the placement of the right foot to find the very best position to do this motion. Notice what the head is doing as you push down with the right foot.

3. Repeat the previous motion using the left foot to push, lifting the left hip. Check that the right leg is resting and relaxed. Position the left foot in different places until you discover the position that allows the most efficient push.

4. Bend both knees and put both feet on the floor while making sure that each leg is independently balanced. Place your arms away from your body. Slowly tilt your knees to the left. Note the shift in pressure across the soles of your feet as the legs tilt. What happens at the neck, chest, and upper back? Does your head want to turn right or left? Where does the work come from to tilt the knees? Notice what happens inside your mouth and throat when the knees tilt. If you feel an urge to turn your head in one direction as you tilt your knees, go ahead and exaggerate that motion.

   Let your legs go a little further each time you tilt them and feel the effect on your shoulders. Rest completely with your arms and legs long.

5. Repeat tilting your legs to the other side as you roll on your pelvis and roll across the bottoms of your feet. Rest.

✓ ***Awareness Advice:***

Do not continue this lesson with the legs crossed if you have had hip joint replacement surgery.

6. Bend both of your knees and place your feet flat on the floor. Cross the left leg over the right at the knees. Balance on the right foot and rest in this position. Open your arms out to your sides away from your body. Gently and slowly tilt your legs to the left and then back to the middle. Remember that you are not trying to accomplish something, but you are trying to feel what you are doing. What movements occur in the spineand the back as you return your legs to the middle? Close your eyes and allow your jaw to rest so your head can move freely.

✓ ***Awareness Advice:***

When you cross your knees, make sure that you feel the back of your left leg fully crossed over your right, as if you are sitting in a chair with crossed legs. Make sure you do the movement gradually so you increase the degree of tilting in small increments.

If crossing one leg over the other at the knees is not easy, try just letting the ankle of one leg rest on the thigh of the other. Balance in that position and rest. If this modification is easy, then proceed with the lesson. If you use this modified leg position, the tilt of the legs should be very small so that you can feel what is going on in the throat and the movement of the head. If you cannot get comfortable with one leg crossed over the other, then just imagine the activity of the remaining segment of the lesson. Try the lesson again later.

7. Repeat the previous activity allowing the right leg to cross over the left.

   Be sure to do the movement so that you can notice which of the two sides is easier for you. As the movement grows larger, make sure it never grows beyond your ability to attend to all of yourself. Does your lower back arch when your legs tilt down? What does your back do to try to influence the movement of the legs? Intentionally take your head in the opposite direction of your legs. What does that do to the movement?

✓ ***Awareness Advice:***

The more we discover and explore new and unusual movements, the more we awaken parts of our bodies that we haven't used in years. Part of what we are after in this method is to involve everything it is possible to involve with our attention and with our movements.

8. Bend your knees and balance the feet on the floor. Put your right hand on your chest and your left hand on your belly. Inhale a small amount of air and hold your breath. Push your breath down into your belly so the belly rises and you feel the motion under your left hand. Then pull the breath up into your chest, feeling the chest rise with your right hand. Go back and forth pushing the air up and down.

Push the air down toward your pelvis and up toward your neck. After you rest, switch the hand position so the right is on the belly and the left is on the chest and repeat the activity. Note if the pelvis is moving with the activity.

9. Cross your right leg over the left again. Tilt your legs over toward your right side and stay in that position. Note how breathing effects your legs and if it rocks the knees.

10. Repeat the previous actions, allowing the left leg to cross over the right. Tilt your legs all the way to the left side and back and notice how far you go this time. Then tilt the legs to the left and stay there. Inhale and push the air back and forth from your chest to your belly. Can you feel what the knees want to do as you push the air back and forth from chest to belly? After you rest enough, get up slowly, stand and notice how your feet contact the floor. How are you balancing? Walk and notice how walking feels to you after the lesson.

MOVEMENT LESSON 10

INTEGRATING THE ACTIONS OF THE WHOLE SPINE

In this lesson, you'll learn to strengthen the muscles of the back while greatly improving flexibility. You will gain the awareness of how to use each part of the spine to help all other parts of the body.

1. Lie on your back, legs long, knees straight, and arms down near your sides. Close your eyes and let your attention drift inside your body. Where do you feel heaviest on the floor? Where do you have the sharpest points of pressure with your back against the floor? Slowly roll your head from side to side. Feel the change in pressure across the back of the skull.

2. Roll onto your stomach with your legs long and easily apart. Put your right palm on the floor with the left on top of it. Bring your face to look toward your left elbow, and rest your right cheek on top of your left hand. If that's uncomfortable, put your hands above your head on the floor and put your cheek on the floor.

3. Bend your knees and tilt your legs to the left so the arch of your right foot massages the inside of your left ankle and leg. Then tilt to the other direction and massage the right leg with the left foot. Rest with your legs long.

✓ *Awareness Advice:*

You'll feel your pelvis starting to turn as you tilt your legs to each side. Make the movement light and playful to heighten the awareness of your feet.

4. Bend the knees again, your face looking to the left, and bend back your ankles, as if you are standing on the ceiling or balancing a book on your feet. Now, lift your left thigh off the floor as you push your foot up to the ceiling. Only go a small amount. Do this several times, as comfortably as you can, and then rest with your legs long.

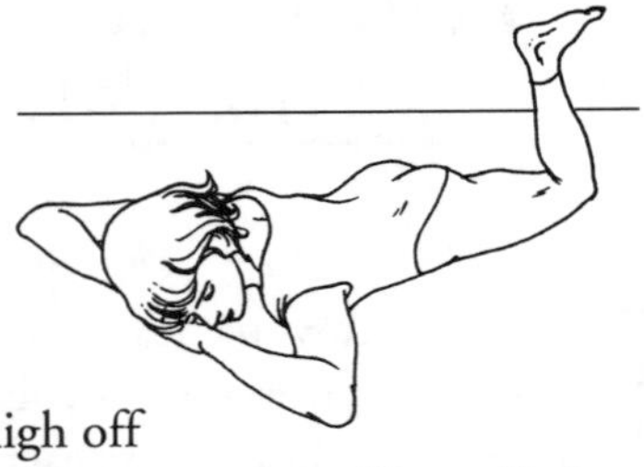

✓ *Awareness Advice:*

As you lift the thigh from the floor, can you feel your breathing and the work throughout your back? Make sure you don't go very far. A slight movement is sufficient.

5. Again, bend the knees to right angles and place the feet as if you had a book on each foot. Turn your face to the right and place your right hand on the back of your left hand, resting your cheek on your right hand. Now, lift your right leg off the floor as if you are pushing the book up to the ceiling. Do this several times and then rest with your legs long.

6. Again, bend the knees and ankles to right angles. Now, place your knees, feet, and ankles together, as if they were glued. You can even imagine that they are tied together with a silk cord.

Tilt both of your legs as one unit to your left side, a little bit, and then back again. Do the movement gradually, paying careful attention that your ankles and knees stay together the entire time. After you do the movement several times, rest completely on your back and feel your back against the floor.

✓ ***Awareness Advice:***

For every inch of slipping or sliding between the two legs, you lose movement in some vital part of your spine. The difficulty with tilting our legs as one unit comes from not knowing how to use the muscles of our back properly, so let the constraint of keeping your legs together create new freedom in your spine.

Also, as you tilt your legs to the left, you'll be rolling over your left thigh, and of course your pelvis will roll as well.

7. Roll back onto your stomach with your left hand on top of your right and look to the left with your cheek on the back of the left hand. Again, bend your knees and ankles to right angles and glue them together in one unit. Slowly tilt your legs to the right and back again. Explore how your pelvis and back help you as you roll over the right thigh. Make sure not to slip. After doing this several times, take a complete rest on your back with your legs long.

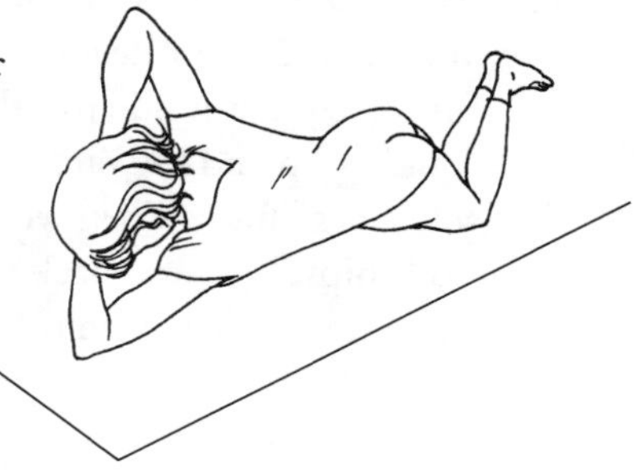

✓ ***Awareness Advice:***

Can you feel the movement pulling on your left shoulder and elbow? If your right shoulder hurts when your legs tilt to the right, simply put your right arm down at your side. This will also give you greater range because your shoulder blade is now out of the way.

8. If your back feels a little tight from finding new ways of using your muscles, give yourself some relief in the following manner: Draw your knees above your chest and hold on with one hand to each knee. Put your knees together and slowly move them in a large circle. Eventually, you can remove your hands from your knees and feel the work of the abdominal muscles and back as your knees go in circles. If you need to, rest, but make sure you go in the other direction with your circle.

9. Once again, lie on your stomach with your left hand on top of your right, looking to your left, knees and ankles glued together at a right angle. Tilt your legs to the left and explore having your face look to the right as well as to the left while doing this. Does it make any difference to the tilting of the legs which way you face?

10. Explore the same movement tilting to the right.

11. With your forehead on the back of your hands, or your face turned in either direction, tilt your legs as one unit, lightly and freely, from side to side. Can you do it with your face looking up and forward as well? Rest on your back with your legs long. When you stand and walk, notice how your hips, legs, and back move, and also if you have more awareness of where your feet are without looking at them.

MOVEMENT LESSON 11

ROLLING LIKE A BEAR

Here is one last movement, which you might want to do when you do not have enough time for anything else. It's one of the most comforting movements to help lower muscle tone and reduce tension throughout the body, in addition to the back.

1. Draw the knees up, above the chest, and hold on with one hand to each knee, below the kneecap. Hold on firmly from the outside and feel the hand contoured to the knees.

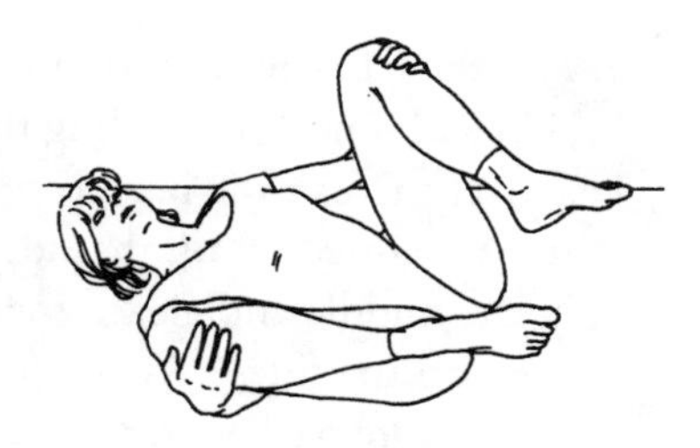

2. Pull the knees wide apart with your arms and push them together again. Imagine you are playing the accordion, as you push and pull your legs apart and together.

✓ ***Awareness Advice:***

Be sure to let your feet fall apart freely when you open your knees.

3. The next time they're wide apart, let them stay that way. Feel the elbows reaching to the floor to help pull them more. Now, with this open feeling, roll to the left by pulling your left leg further to the left side and also rolling your head to the left. Your entire body should now be

lying on its left side. Close your right knee down to your left knee, as if you're bringing your accordion together. Remain holding on to your knees and rest on your left side for a moment.

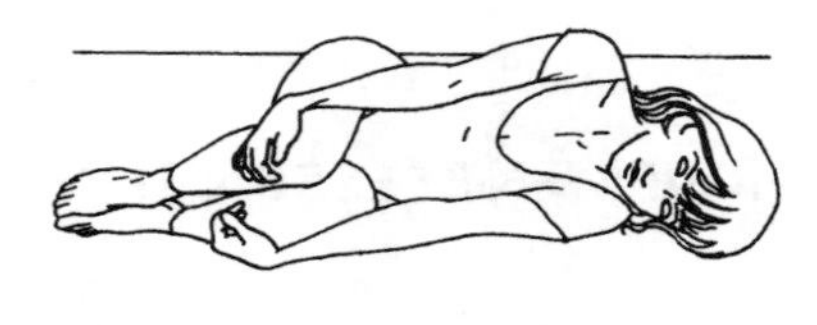

4. Pull your right knee up to the ceiling, across your chest, as far as it can go, and begin rolling to the right side. Let your head turn to the right. Keep on going as far as you can comfortably go keeping your legs wide apart. When you lie on your right side, close your left knee down on top of the right knee. Think of taking a one-second nap lying on your right side, still holding onto the knees but completely at rest.

5. Pull the left knee across to the other side. Roll from side to side with one knee leading the movement. One knee pulls actively across, the head constantly moves. See that you remain curled up on the sides, so you're open to the ceiling in the middle and closed together, almost in a fetal position, on the sides. As you feel one leg actively pulling, feel your entire back getting a massage from the floor.

6. Feel the head now. Notice that it turns. Can you keep that turning even? See that the skull is always moving. It never stops until you're fully on the side. If you allow the head to roll fully, you'll find yourself looking down into the floor when you're folded on your side. See that the head never accelerates or decelerates. It's a constant, steady movement. Allow the whole back to spread itself across the floor, the ribs, the lower back, the pelvis, and the head. Move at any speed you find comfortable, as long as you know that your head leads the movement.

Leave it and take a full rest on your back. Let your arms and legs go long and see what your body feels like now. Feel the quietness, the fullness and ease in the breath and in the movement of the ribs.

✓ ***Awareness Advice:***

Remember that good back exercise is good exercise for your whole body. Continue to lie here enjoying the pleasantness of your sensations. When you feel ready, stand and walk, continuing to enjoy the way you feel.

SECTION V

THE JOY OF WALKING

Many people find that walking is the most enjoyable form of natural exercise. Not to mention the fact that it is essential to get from place to place. However, most people don't walk well and they don't even know that they don't walk well. Many people simply don't enjoy walking or they find that when they have to walk faster to get somewhere, it becomes unpleasant. In certain situations people find their balance to be not very good or feel that their hips or knees occasionally ache or feel stiff. In this section you will learn how to find much better balance than you've ever known before and you will have much smoother actions of hips, knees, ankles, and feet. These lessons will not only save your feet from stress if you walk but are also very good lessons for runners. Finally, you will have a more elegant and graceful walk that will give you the ability to more fully enjoy one of life's most basic activities.

MOVEMENT LESSON 1

IMPROVING BALANCE

Many people don't even realize that they have difficulties balancing because they have never been diagnosed as having a balance disorder. However, a startling number avoid situations where their balance could be challenged, and frequently find themselves looking down at the floor or sidewalk whenever they move about. This lesson will provide an opportunity to greatly improve your balance and contribute to your walking enjoyment.

✓ ***Awareness Advice:***

This lesson will be much easier to perform if you are standing behind a chair or near a wall that you can touch to provide support. This lesson will challenge your balance, so make sure you feel comfortable and safe.

1. Cross your right leg over your left so that the ankles are crossed and your right foot stands on the floor just outside your left foot. Make sure the whole foot is on the ground and both heels are down.

2. Slowly shift your hips left and right as far as it is comfortable to go. Reduce the effort you make to maintain balance. Do this enough times so that you learn to feel secure and comfortable, and have your jaw relaxed and your eyes looking out into the room, breathing fully. Uncross your legs and rest while standing.

✓ *Awareness Advice:*

As you shift your hips you will feel yourself pressing more on one foot, and then on the other. As your hips move from side to side, your head remains stationary.

3. Now cross your left foot in front of the right, so your left foot stands on the floor just outside your right foot. Again, oscillate the pelvis from side to side until you feel certain of how to organize your back, hips, head, neck, and shoulders, and until this feels like a natural movement. Uncross your legs and rest while standing.

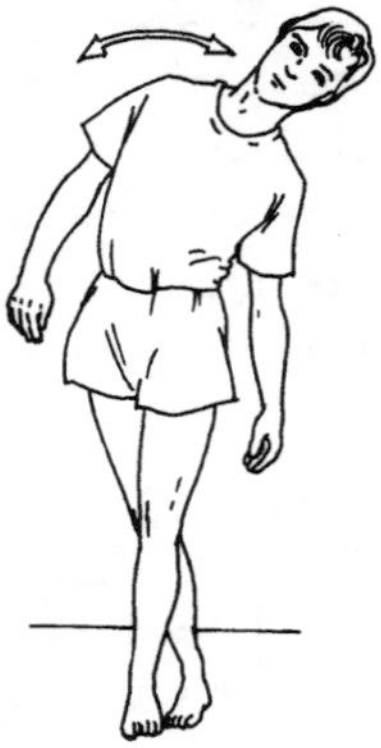

4. Cross the right foot over the left again as in step 1. Now lean your head and shoulders from side to side. What do your hips need to do? Learn to use all parts of your body to make it comfortable. Gauge your comfort by the ease of your breathing. Uncross your legs and rest while standing.

5. Cross your left foot over the right. Again, lean your head and shoulders from side to side. Is it easier or harder with your legs crossed in this arrangement? Uncross your legs and rest while standing. After you have rested, walk around the room. Do you notice your hips more clearly?

6. Return to the chair or wall for some support. Once again, cross your right foot over the left so that the ankles are crossed and the right foot stands outside of the left. This time, move your pelvis forward and backward so you press toward the balls of your feet and then toward the heels. If you are comfortable enough with the movement you might be able to lift both heels off the floor when you go forward with the pelvis. Perhaps you can also lift your toes and the balls of your feet when you push your pelvis backward. Uncross your legs and rest while standing.

✓ ***Awareness Advice:***

As you move your pelvis forward and backward, let your arms, shoulders, and neck relax completely, so you can feel your head and your arms moving in opposition to your pelvis to help balance you. Your head will feel as if it's reaching forward and pulling backward.

7. Cross your left foot over your right so the ankles are crossed with the left foot just outside of the right. Oscillate the pelvis forward and back as in step 6. Uncross your legs, rest, and then go for a short walk around the room.

8. Place one of your hands on the back of your chair or on the wall and again cross your right leg over your left. Stand tall, but comfortably. Turn your head to look around the room. As you do, look in back of yourself as far as possible. Look at the ceiling and the floor. Can you feel the adjustments your hips need to make?

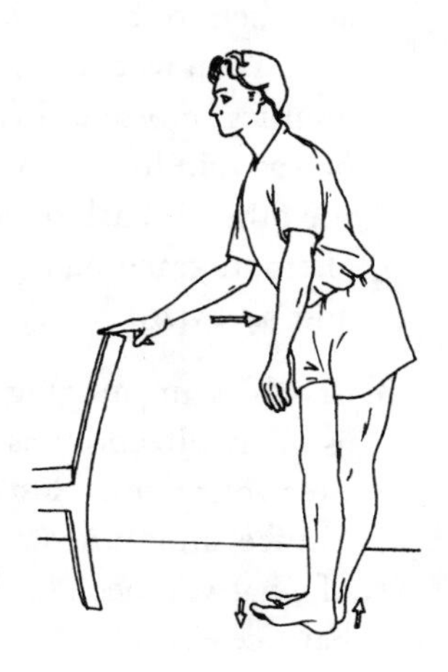

9. Recross your legs with the left one crossed in front. Continue looking around the room in this arrangement. Which way of crossing the legs gives you better balance? Rest standing comfortably.

10. Stand with the right leg crossed over the left. Now make a circle of pressure on the bottoms of your feet. Move your pelvis in a circle so you feel as if both your pelvis and the bottoms of your feet are tracing circles, one in the air and one on the floor. Uncross your legs and rest.

11. Cross the left over the right and explore your circle of pressure this way. Which side creates the roundest, smoothest circles on the bottoms of your feet? Rest by walking around the room. You might find your legs can walk in a narrower path then usually.

12. Once again, cross the right leg over the left and raise both of your arms out to your sides at the height of your shoulders. Imagine holding on to two swords and lunge first to one side, and then to the other, as far as you can reach, keeping your legs crossed. Reach one arm in front of you and the other in back of you and alternate reaching forward and backward with your arms.

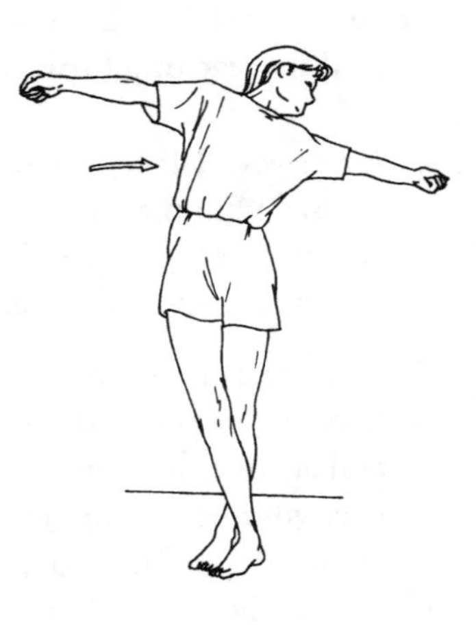

13. Try the same lunging and reaching actions of your arm in as many directions as possible with your left leg crossed over your right. Rest by going for a walk and notice if you can feel an improvement in your balance and an awareness of what you need to do with your hips to help yourself balance easier.

MOVEMENT LESSON 2

DISCOVERING YOUR FEET AND ANKLES

Many people don't realize the exact location of their feet and ankles as soon as they are in an unusual position, or when they're faced with walking on uneven or uncertain terrain. That's why you'll see them look down at their feet whenever they walk in unfamiliar surroundings.

In this lesson you'll learn to heighten your sensation of your feet and ankles, which will increase your ability to balance, walk more steadily, and as many people have found, even to run more easily.

1. Lie on your back, arms and legs long, with your hands at your sides. With your eyes closed, notice how your legs lie on the floor. Does one seem closer to the floor? In your mind, can you measure the distance from the heel to the tip of the big toe on each of your feet? Which feels longest? Which foot feels widest? Are the toes pointing at the same angle, or is one pointing more toward the ceiling than the other one?

2. Roll over and lie on your stomach as comfortably as possible. Experiment with different ways of placing your arms and head on the floor until you find what's most comfortable. Bend

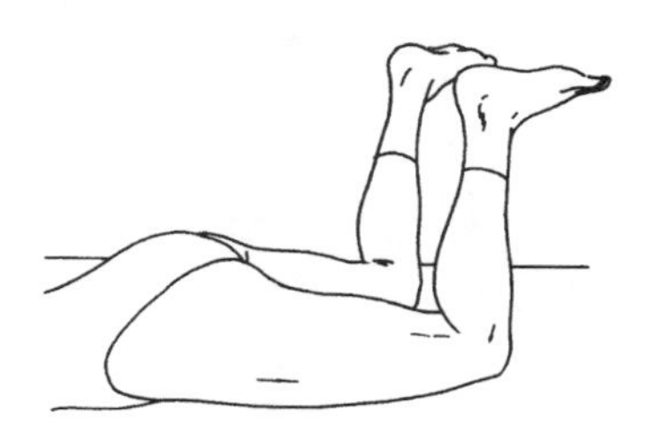

your knees to make right angles at your knees and ankles. Can you feel the angle of your knees and ankles when you aren't looking at them?

3. Point your feet toward the ceiling and then flex them back toward the floor. Keep your knees bent and see that your toes don't move; the motion is in the ankle, and the toes merely follow. Gradually decrease the amplitude of the movement so the feet come to rest in a neutral position.

✓ *Awareness Advice:*

Most people feel that their feet are in a neutral position when they are actually very extended at the ankle because they are used to tightening their calf muscles, more than the muscles in the front. They don't realize they have an unnecessarily high muscle tone.

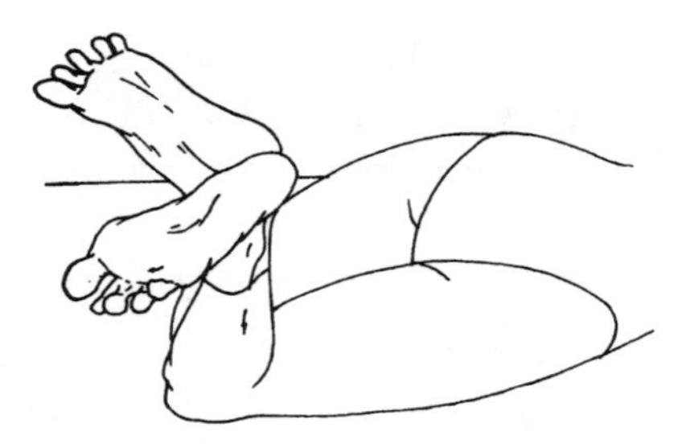

4. Join your bent knees together along with heels and balls of your feet. While keeping your knees and heels together, separate the toes and balls of your feet from each other. Turn both feet so your toes point outward and then come back together several times. Rest with your legs down.

✓*Awareness Advice:*

The feet do not move toward the ceiling or the floor. Only the toes point away and together. You'll almost feel like you're pivoting on your heels. Do the movement until what you're doing becomes clear and smooth.

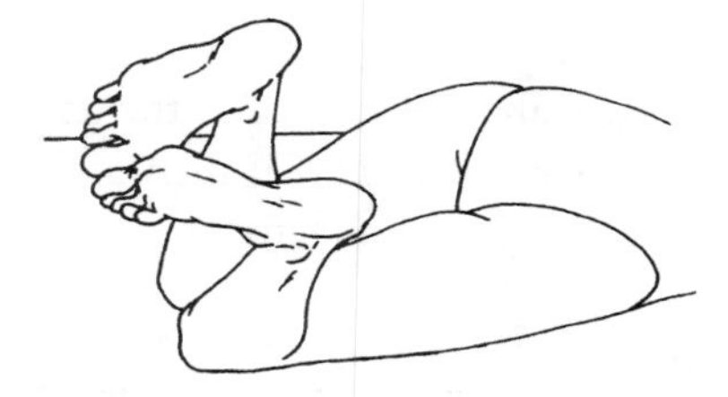

5. Bend the knees again with the knees and ankles together. This time, keep your toes together and separate only your heels, again, without flexing or extending your ankles. Is it easier than separating the toes? Rest with your legs down.

6. Again, bend your knees and put the legs together. Alternate separating the heels and then the toes. Do it with your head turned in different positions. Rest.

7. Again, with the knees bent and the legs joined, move only the toes of the right foot away from the left foot so only the front of the right foot moves, and the heels stay together. Pause, then bring only the right heel away from the other heel so the toes stay together. Alternately swing your toes and then heel of the right foot away from the stationary left foot.

Pause, and do the same thing with your left foot, keeping your right one stationary. Rest with your legs long.

8. Bend only your right knee and right ankle in 90° angles, and turn your face to the right. Keep your heel exactly where it is and point your toes further to the right, then back in. It's the same movement as before, except you don't have the other foot to rest against. Can you do it without straining in your neck or face?

9. Now keep your big toe exactly where it is and take your heel out and back in, as though someone were holding on to your big toe. Make sure the foot stays parallel to the ceiling. Rest.

10. Bend your left knee and ankle, face to the left, and repeat the movements in steps 8 and 9. Which of the two feet is more clear when doing this movement?

✓ ***Awareness Advice:***

Some people find that one foot feels easy to control, and the image of the foot's position is sharply focused. Sometimes one or both feet may feel vague and the movements may be difficult to control.

11. Please stand in front of a wall. Face the wall, standing a small distance away, supporting yourself with your hands. Make your feet parallel and a little bit apart. Turn the front of your right foot further out to the right and then inward, so you are pivoting on your heel.

✓ ***Awareness Advice:***

The action is in your hip joint so that only your foot moves, and your pelvis does not rotate. Move slowly and steadily so the foot glides over the floor evenly. Don't stiffen yourself or lean into your hands against the wall. Your hands are only there to help you track your upper body so that it doesn't turn while you turn your foot. Your interest is in differentiating the foot and hip.

12. Now stop in a neutral position, and begin moving the heel in and out, pivoting on the ball of your right foot.

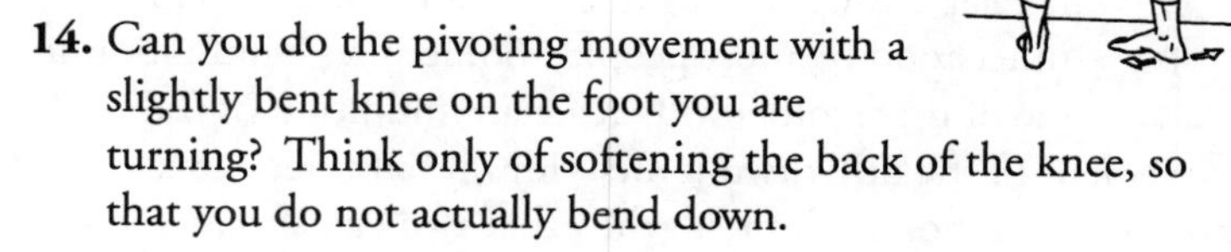

13. Pivot on your left heel as your left toes point outward and inward, and then pivot on the ball of your left foot as your heel goes outward and inward as you did on the other foot. Which of the two legs feels more natural when you do these movements?

14. Can you do the pivoting movement with a slightly bent knee on the foot you are turning? Think only of softening the back of the knee, so that you do not actually bend down.

15. Walk around and feel if you have a heightened awareness of your feet, knees, and hips.

MOVEMENT LESSON 3

FREEING STIFF HIPS AND KNEES

The relationship of hips to ankles and knees is poorly understood by most people. Sometimes a limitation in one of these joints will create an inhibition of the movement in the other two joints. Ideally, all three move together, and if one is limited, the other two can simply move a little more. Increasing mobility involves learning to use them differently, and also learning to balance better.

✓ ***Awareness Advice:***

You will need a chair with a back high enough so that you can stand next to it and comfortably place your fingers or hand on it without having to reach down or bend over.

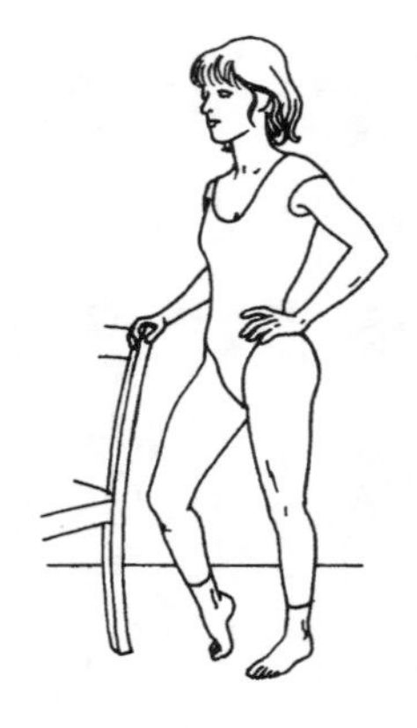

1. Face the back of the chair and stand off to the left side so you have room in front of you, with your right hand on the back of the chair. Bend your right knee just enough so your right heel lifts off the floor. Repeat the motion several times. Do not lean heavily on your hand.
2. Now lift the ball of your foot from the floor so all the weight rocks back to the right heel. The left leg is stable all the time. Check that you can breathe and do the motion at the same time.

✓ ***Awareness Advice:***

When the heel comes off the floor, can you feel that the right knee simply moves forward lightly and the hip softly folds? To lift the ball of the foot, feel what's necessary to do on your stable left leg and with your back and pelvis. Be sure to keep your upper body easily upright so that you can look out and around.

3. Now rock back and forth from the heel to the ball of the right foot. Your whole body does not need to move forward and backward, but you might find your right hip moving forward and backward slightly. Be sure not to exert pressure on the chair. Walk around for a moment and notice any differences in your two legs and feet while walking.

4. Return to your chair and place your left hand on the back while you stand to the right of the chair. As you did on the other side, first lift the left heel off the floor and observe how you can do it lightly and easily. Can you feel the work of stabilizing yourself in the right hip?

5. Now lift the ball of the left foot from the floor. It might be easier or harder than on the right side.

6. Now rock from the ball to the heel on the left foot, making sure that your upper body feels free and your eyes can look up and out. Every so often close your eyes to deepen the feeling, and then open them to link looking out. Rest by walking and feel what's happened to the motion of your legs and to your balance.

7. Return to the left side of your chair with your right hand supporting yourself on the back of it. Alternately lift the ball of one foot and the heel of the other. So, for example, as the ball of the right foot lifts from the floor and you press on the right heel, at the same time your left foot will pick up as you stand on the ball.

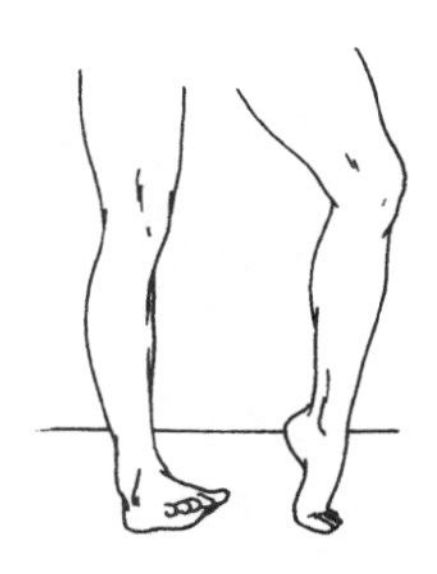

Make it a smooth and steady alternation. Do the movement very slowly at first and then learn to go quickly. Master all speeds.

✓ ***Awareness Advice:***

Can you do this movement with your eyes closed, making sure that you're balanced and breathing, and not gripping your chair? Then, can you do it with your eyes open while looking to both sides, to the ceiling, and actually focus on something in your room or out your window? Can you keep the movement going steadily whether the eyes are open or closed? If you feel safe enough, take your hand away from the back of the chair.

You will probably feel your hips shifting from side to side. Let that happen.

Rest and enjoy your new ability to walk more easily.

MOVEMENT LESSON 4

INCREASING PROPULSION

Most people are unaware of how they can increase their power and speed in walking because they are not aware of how they can propel themselves as they push off with their rear leg. In this lesson, you'll learn how to improve your push off as well as your pelvic rotation, which will increase the length of your stride and put more power in your walk. When you're done with this lesson, make sure you bother to open the door instead of just walking through it!

1. Roll over onto your stomach. Rest your arms with bent elbows so your palms are on the floor somewhere near your head. Let your legs be a comfortable distance apart. Notice which way your heels go. Do they point inside or outside compared to your toes?

   Is there more weight on the right hip or the left hip? Do you have more space under the right shoulder or the left shoulder?

2. With your face to the right, pull back your right foot at your ankle and stand your toes on the floor. Imagine that the balls of your feet are against a wall. Push your toes into the floor as though you were pushing them into your imaginary starting block or wall and observe what happens to the force that goes through your leg from the pushing foot. Can you push so that each time you push with your right foot, the knee lifts from the floor and becomes straight?

✓ ***Awareness Advice:***

Every so often, you may have to pull the toes back to keep them standing solidly. Wearing socks for this lesson will get in the way, so make sure you take them off.

As you push with your foot, you'll feel the back of the knee heading toward the ceiling. Make sure you set the knee down and relax completely at the end of each push, even though you remain standing on the toes.

3. The next time you push through the foot and straighten the knee, could you push further so that the right side of the pelvis begins to raise off the floor as it tilts to the left?

   You'll feel your pelvis rolling further to the left as the right side picks up, and as you begin to feel your back participating in the movement. Rest with your ankles straight.

✓ ***Awareness Advice:***

Make sure you don't stiffen your body as you push through the right side. The other side of you is completely at rest. Each time you release the pushing foot, make sure your pelvis and knee rest on the floor. As you do the movement, can you feel your ribs moving? Can you feel the push go all the way up your shoulders and into your neck?

4. Bend back the right ankle to stand on the toes. What toes are you standing on? Your big toe? Several toes? Rock your heel more to the outside so you're standing on the little toes of your right foot. Push again into the floor to lift your knee, your pelvis, and ribs on the right side of

your body. Observe how much more easily the force moves into your upper body. Rest with your ankles straight.

5. Now stand only on your big toe so your heel turns inward. Don't use any other part of your foot at all. Again, push to lift your knee and thrust all through your body. Observe how much power you have in this position compared to when you pushed just with your little toes. Rest.

6. Now choose another place to stand on your toes. Explore the part of your toes that would give you the most power, so that for the least effort you get the most motion through your upper body. Roll over on your back and rest.

✓ ***Awareness Advice:***

There is a place on your foot where the bones of the legs, pelvis, spine, and ribs move the easiest with the least amount of muscular effort. Most people don't discover where this place is, even though they walk around their entire life.

Slowly come to standing and feel the difference in your weight bearing on the two legs. Walk around for a while and feel the difference in the two sides of your body. Which leg feels most secure and comfortable? Try pushing off on the ball of your right foot, and then on the ball of your left. Can you feel the difference?

7. Lie on your stomach again with the arms in an easy triangle and turn your face to the left. Repeat steps 1-6 slowly and carefully on the left side. After you stand on your left toes, don't push as far as you can right away. Go slowly enough that you can feel how you do the movement and how you breathe as you do the movement.

✓ ***Awareness Advice:***

Learn to let the movement become easier each time you push. Make sure that you can distinguish which of your two legs makes an easier connection from foot to head. Occasionally do the other side again so you can feel the difference in your effectiveness as you push.

8. Now pull back both feet and stand on both sets of toes. Again, find the place that feels most secure and push with both feet to lift both knees. See that you can lift your knees from the floor as you push. Then let the knees down and each time you push with your toes into the floor against that imaginary wall, see that the force can be driven up through your legs so that the knees come off the ground. Rest with both ankles long.

9. Bend back both ankles again and push not only to lift the knees, but to lift your pelvis from the floor as well. You'll have to use your back to help you do this. So push with the feet until the knees straighten and the bottom of the pelvis lifts from the floor. Feel as though you are peeling the lower parts of your body off the floor. Can you peel your belly and some of your ribs off as well? Go up and down slowly, several times. Learn how to harmonize the thrusting through the feet and the straight knees with the lifting of the spine so that the movement feels light, easy, and steady.

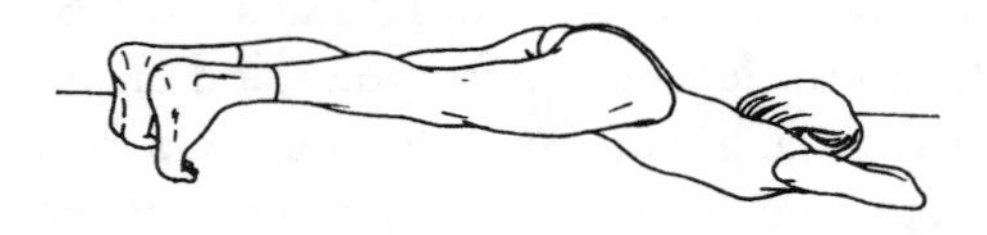

Rest once you feel that you can make the movement smooth and easy.

✓ *Awareness Advice:*

As you are pushing, see that you take the time to let your knees back down to rest on the floor after your pelvis settles. You may have to remember to pull back your feet and ankles occasionally so you can stand on them strongly.

People do all sorts of exercises trying to run farther or faster, jump higher, etc. Rarely do we learn to pay attention to how the muscle force in our foot transfers to the thighs and into the pelvis and back. Often, people will waste hours and a lot of energy trying to make their muscles thicker and stronger to improve what they're doing. But thick, strong thighs won't help you walk easier or jump higher. The legs of high jumpers are no thicker than anyone else's. Effective propulsion is a skill that requires that the movement passing through our skeleton and the muscles that move the skeleton all work together in a coordinated manner to propel the body in the direction we intend to move.

MOVEMENT LESSON 5

THE SHOE-SAVER LESSON

Walking is a habit. People continuously wear out one part of their shoes more than the other parts because they are not aware of what part of the foot they use and overuse and what parts of the foot are underused. In fact, many need a new pair of shoes, not because the bottom is worn out altogether, but because only one place on the shoe is worn through. In this lesson, you will learn to walk with many parts of your feet and discover the relationship of different parts of your feet to the way you habitually use your knees, hips, back, and even your upper body.

1. Start by simply walking around the room. Notice: does one of your arms swing more than the other one? Can you identify which part of your feet you walk on the most, i.e., inside, outside?

2. Stand with your feet a comfortable distance apart, close your eyes, let both arms hang down freely, and feel which leg bears more weight. Which foot seems closer to the ground? Are both feet contacting the floor in the same way? Or, do you stand more on the front of one foot and the rear of the other, or on the outside edge of one foot more than the other?

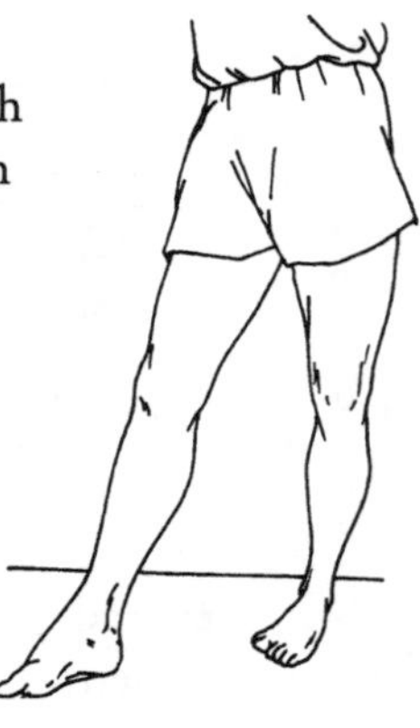

3. Lift the arch of your foot from the floor so you roll up to the outside

edges of your feet. As you do this several times, observe what happens in the rest of your body, especially your back, hips, and shoulders. Remain standing on the outside edges of your feet.

Walk around only on the outside edges of your feet for a while and feel what happens to your upper body, especially your back and belly, and the position of your arms, and the tension in your neck. Rest standing normally. Feel the highlight of the impression your feet make on the floor. Has it changed?

✓ ***Awareness Advice:***

If you feel like lying down or sitting and resting for a moment, please do so. Otherwise, rest in a standing position. Do not fatigue yourself; tired muscles will not learn a new way of working.

4. Make sure your arms are hanging freely and your legs are a comfortable distance apart. Can you lift the outside edges of your feet from the floor? What must happen throughout your body as you do this? Remain standing only on the inside of your foot, with the outside edge lifted. Walk on the inside edge of your foot and observe the consequences throughout your body. Rest in any position you want, but take a moment to simply stand and observe how your feet contact the floor now.

✓ ***Awareness Advice:***

For some people there are obvious gross involvements of their arm position, head, back, breathing, and so on. For others, the changes in the body aligned with changes in the foot position are more subtle. In either case, notice not only the changes in position and muscle tension in the trunk, but be aware of what happens to your throat, your eyes, and your breathing as well. Some people find that their peripheral vision expands or narrows, depending on which side of the foot they walk on.

5. Pick up the balls of your feet so you rock back onto your heels. Do this several times until it's clear what you need to do to perform this action. Remain standing only on your heels for a moment and then go for a "heels-only" walk. What does this require of the rest of your body. If it's too difficult, you can allow the front of your foot to touch the floor a little bit to help your balance. Rest while standing and feel your imprint on the floor.

6. Now, lift your heels off the floor to stand on the balls of your feet. As you do this several times, feel what happens to your chest and the orientation of your face. Walk around on the balls of your feet only and observe the consequences. Rest and feel your impression on the floor.

✓ ***Awareness Advice:***

Which of these four positions of the feet that you've practiced thus far would be the most difficult for walking? And which would be the easiest? You might find that one of these exaggerated positions of the feetis, in fact, normal for you and doesn't feel very exaggerated at all. This reveals your habitual way of organizing your feet, ankles, knees, hips, and back in support of your habitual pressure on your feet. You might want to look at your shoes to see if they are worn-out in a place consistent with your easiest foot position for walking.

7. Now squeeze your toes as if grasping the floor and walk with your toes curled under. What other parts of your body come under stress besides your toes? Rest while standing.

8. Now lift your toes and point them to the ceiling and walk without the toes touching the floor. Observe the pattern of use throughout your body. Rest.

9. Turn both of your feet out and walk leading with the inside of your heel. What does this do to your pelvis, back, and shoulders? Is this comfortable or awkward? Rest. Now turn your legs, toes in, heels out, and walk this way. How does this compare in the organization of your trunk, with having your toes point out? Rest.

10. Just walk normally for a while and feel how many options you have in the use of your feet against the ground, as well as all the different ways your upper body can support whatever pattern of pressure you choose on your feet.

✓ ***Awareness Advice:***

As with all the lessons in this book, the intention here is not to insist on the right way to walk, but first, to make you aware of your habits of walking, and second, to allow you to experience several options so that walking can be done in many ways. You can discover which way or ways are best for you. This is a more significant achievement than imposing some unfelt pattern of body use upon ourselves.

MOVEMENT LESSON 6

CREATING AN ELEGANT WALK

When we walk, our shoulders and hips need to move in opposition to each other as do our elbows and knees and our hands and feet. The more we reduce this counter rotation of the upper and lower body, the stiffer and more awkward our way of walking becomes. In this lesson you will learn how to move your legs more easily by using your arms and shoulders.

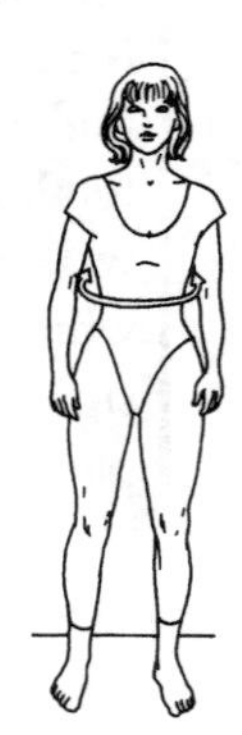

1. Stand with your feet a comfortable distance apart. Slowly and carefully turn your chest and shoulders from side to side without moving your pelvis at all. You might need to look at a mirror to make sure you are really able to do this. Do not try to make the movement large, instead work to make it distinct and clearly felt.

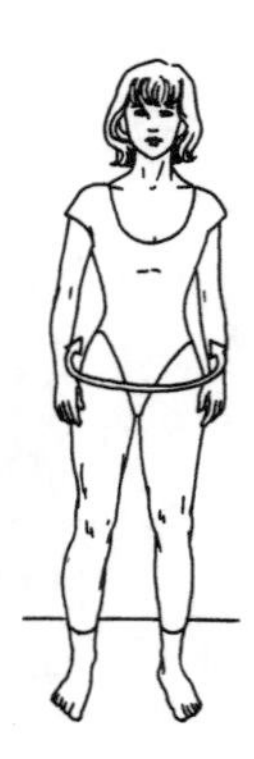

2. Now, turn your pelvis from side to side keeping your chest and shoulders stationary. Again, your attention is on the clarity and simplicity; the size of the movement is unimportant. Rest while standing.

3. Stand comfortably with your feet apart as if you were preparing to walk. Bend your elbows until your forearms are parallel to the floor, and make fists with your hands. Soften your knees so they are not locked straight and

slide your right fist forward as you take your left elbow backward. Then alternate so that your left fist pushes forward and your right elbow pulls backward. Alternate many times.

Rest while standing or walking.

✓ ***Awareness Advice:***

As you move your arms, pump them forward and backward. Let them swing freely so that your shoulders move forward and backward and you feel your chest turning from side to side. As you do this, can you feel the result in your pelvis? Is it stable or could you allow it to move slightly so it turns opposite the directions of your shoulders and chest. Let the swinging of the arms become larger until you feel the effect on your waist and pelvis.

4. Now take your right hand and place it on the side of your right thigh near the hip keeping your elbow straight. Do the same with your left hand on your left thigh. Go for a walk so that your arms weld your hips and shoulders together to eliminate any and all counter rotation in your body.

5. Can you walk with your right hand on your right thigh and your left arm swinging freely? Because most people move one arm and shoulder more than the other, you may be exaggerating a habit you already have.

Release the right hand and walk normally.

✓ ***Awareness Advice:***

Be sure that you feel how the right hip and right shoulder move forward at the same time.

6. Put your left hand against your thigh and move the hip and shoulder together on this side. It might feel more awkward or more coordinated than the other side. Release the left hand and walk normally.

7. Walk exaggerating the forward and backward movement of your shoulders. Try to keep your head more or less centered as you do this, and allow your arms a larger amplitude. Notice what effect this has on your hips and legs.

8. Now, exaggerate the motion of your pelvis as it turns forward to follow each leg. You might feel your shoulders still moving in the exaggerated manner also. Now, sense the motion of your knees as you walk, and exaggerate the forward movement as you walk. Think of aiming your knees toward someplace you want to reach, and let your knees reach for that place as you go.

9. Walk normally for a while and observe if you have more awareness of how your trunk, shoulders, and legs work in harmony when you walk. As you're walking, play with exaggerating your arm swing for a while, then your shoulder movement. Occasionally, exaggerate your hip rotation for a bit, and sometimes exaggerate the forward reaching motion of your knees. Experiment with all these possibilities to have a more graceful walk.

✓ ***Awareness Advice:***

The more parts of your body you move as you walk, the more graceful your walk will be. The fewer parts that move in response to other parts of your body when you walk, the more effort, resistance, and awkwardness will appear in your walk. When you begin to move many parts of your body fully as you walk, it may feel exaggerated. But don't worry, nobody will notice. And if they do, all they'll see is that you look more elegant and graceful.

SECTION VI

The Physical Act of Thinking

The mind works very much like a muscle. In order to think, or even to focus your attention, muscular contractions must take place in the pupils of your eyes and in the muscles that move your eyeballs and face. Even the way you breathe changes as you begin to concentrate. Without this muscular cooperation, you would not be able to think.

For many people thinking is more tiring than exercising. In fact, many people feel more clearheaded and are able to read, write, or think for longer periods of time after exercising. So, we know that our muscles and organs support our cognitive abilities. However, temporarily relieving muscles of stress does not mean that we have significantly improved the way we use our bodies when we think. Inevitably, old habits of thinking creep back into the "thinking muscles" of our eyes, mouth, and neck.

Many people don't realize that in order to read we must subvocalize, i.e., speak the words that we are reading by moving our tongue soundlessly and so quietly that it is unfelt. The speed of reading is tied to this subvocal speaking speed.

When we count or add and subtract, our eyes move. If you think of the number of buttons on your favorite shirt or blouse and count them in your mind, notice what your eyes do. Could you count the stairs by your home or office? Can you feel what your eyes do as you count? (It is easier to feel this with your eyes closed.) Or count the number of folds in a curtain. Can you feel the stopping and starting movements of your eyes at each fold?

How well coordinated these muscle groups of face, eyes, tongue and breathing are determines the physiological basis for thinking skills. In this section, we are going to improve the quality of muscular use that underlies thinking.

MOVEMENT LESSON 1

WAKING UP THE TONGUE AND MOUTH

In this lesson, you will learn to feel how your tongue is used as you silently read. You will learn to speed up the motions of your tongue, lips and jaw in order to increase the speed of these quiet, unfelt subvocal movements.

1. Find something to read, preferably something very easy and familiar. Sit somewhere where you can relax completely, and begin reading. Can you feel small and subtle movements of your tongue, mostly the back of the tongue? If not, stick your tongue far out and close your teeth on the tongue. Continue to read, and now it might be easier to feel the attempts of the tongue to work while you read. If this is still unfelt, pick up something to read that is difficult and unfamiliar. Then, you will feel the tongue working harder and more vigorously. Rest.
2. Set your reading material down and rapidly move your tongue from side to side inside your mouth by breathing steadily. It helps if you move your tongue back away from your teeth.
3. Touch the tip of your tongue to your palate and move the tip forward to your teeth and backward toward your throat. Begin to develop a very fast flicking rhythm as the tongue goes forward and backward. Rest.
4. Speak in glossolalia. Rapidly utter nonsense syllables as if you were trying to make up your own language. See if you can develop unusual sounds and syllables that are not used

in your native language. You would probably want to make sure that other people are not around. It is best to speak quickly and vociferously. Do it for several minutes.

✓ *Awareness Advice:*

It is easy to feel too inhibited to speak glossolalia, but children do it all the time, and in the play of doing so they develop the articulating movements required for speech. Ordinary tensions throughout our body affect our tongue and mouth as much as our back or shoulders and neck; it is possible to have a rigid and fixated set of muscular contractions when we speak or read. So, let yourself go. Utter complete nonsense as fast as you can, and break those rigid habits of use in the tongue and mouth. Some people say it is easier to do if they let their head move freely as well.

5. Now find some familiar and easy reading material again and read out loud as fast as you can. Challenge yourself to break the "reading barrier." Read at least as quickly as you could perform glossolalia. Do it for at least several minutes.

6. Now find some reading material that you are interested in, but that is unfamiliar to you. Read the material silently as fast as you would like and see if your tongue can now keep up.

✓ *Awareness Advice:*

Repeating this lesson on a regular basis can help maintain faster reading speed. There are children and adults who have doubled and even tripled their reading speed by melting the rigid habits of their tongue.

MOVEMENT LESSON 2

ALTERING USE OF THE EYES TO IMPROVE MATHEMATICAL SKILL, I

Most people form images when computing mathematically, but how we use our eyes to form these images remains hidden from our senses. As I described in the introduction to this section of lessons, whenever we try to count something, our eyes stop at each item that we are counting. This can make even simple addition very slow if the eyes have the habit of moving very slowly and prolonging the duration of the stop. People who can add and subtract quickly and otherwise manipulate numbers skillfully make almost imperceptibly smooth and brief stoppages of their eyes when they count.

Those who are most skillful at mathematics see several items or even "sets" of items at once. The following exercise will help make you aware of how you use your eyes to compute and will also help to give you an ability to manipulate larger numbers or items at a faster speed.

1. Sit or lie in a comfortable manner. It is important that you be as relaxed as possible to reduce muscular stress that would make it more difficult to observe the very subtle contractions of the muscles of your eyes.
2. You will need to close your eyes during this lesson. With your eyes closed, observe any movement your eyes make under your lids. Look at one imaginary coin (or any object you prefer if you don't like to look at coins). Imagine another coin next to that coin. Have the coins be a good distance apart. Notice if it is possible for you to see both

coins without your eyes moving back and forth to each one.

3. Now, move the two coins to the left side of your field of vision, and add three coins to the right side. Can you feel the movements your eyes make to see three coins? Can you try to see all five coins at once, each one distinctly without your eyes moving? If not, remove one or more coins until you can see them all individually without moving your eyes. Open your eyes and remove all your coins. Pause.

4. Close your eyes again and try increasing the number of coins in your field of vision to six. Can you see six coins without moving your eyes to each one? If not, remove one or more coins until you can see them all without your eyes moving to them. Can you see seven without moving your eyes? Rest your eyes.

5. Gradually increase the number of coins or other items in your field of vision one at a time without moving your eyes to see the individual coins. Some people can work their way up to 20 or more.

✓ ***Awareness Advice:***

It takes tremendous concentration to hold several items in our mind without moving our eyes to each one. It might require a little practice every day to increase the number of items in your field of vision.

6. Now can you try taking five items on the left side of your visual field and seven items on the right side and move them together to form 12 items? Be sure to take your time to let 12 items emerge in your visual field.

✓ ***Awareness Advice:***

This kind of concentration can make us hold our breath, squeeze our face, and grind our teeth. Thinking is a whole-body activity. Be sure to take an

opportunity to observe what you are doing with your hands, feet, and breath while you concentrate.

7. Can you keep your eyes stable while you see thirteen items in your visual field? Keep breathing and relax your face and throat while you try. If you succeed, divide your thirteen coins into groups of six and seven, and move them to different sides of the visual field. Can you feel your eyes moving the groups of coins? Now try to do this again, but don't move your eyes while you separate your thirteen coins into two groups. Rest and slowly roll your head from side to side.

8. Gradually keep adding coins or other items to your visual field without moving your eyes while keeping your face, jaw, and hands relaxed. If you can hold a high number of objects in your visual field, can you count them one at a time without moving your eyes?

✓ ***Awareness Advice:***

Mastering this lesson will give you the concentration and focusing skills necessary to compute more rapidly by eliminating unnecessary eye movements when we count. As our eye muscles become more efficient, so will our brain. Many people have never learned how to focus their mind to be able to think more clearly, compute more readily, and read more easily. Mastering this lesson is like a meditation that provides the foundations for further skills. The next requires that you have mastered this reasonably well. By the way, if you can only hold seven or eight or nine coins in your visual field, that's okay, and in fact, it reveals a greater necessity to learn the skills this lesson offers.

MOVEMENT LESSON 3

Altering Use of the Eyes to Improve Mathematical Skill, II

This lesson requires that you have performed and practiced Lesson 2.

1. Again, lie or sit in a very comfortable manner. Close your eyes and place three objects in your visual field. Without moving your eyes, move those objects to the upper left quadrant of your visual field. Now, add four objects to the upper right quadrant of your visual field. Without moving your eyes, reverse the locations of your objects. (You now should have four objects in your upper left quadrant and three in the upper right quadrant.)

2. Also, place two objects in the lower left quadrant and three in the lower right quadrant. Without moving your eyes, can you add the sum of all your quadrants? (If this is too difficult to do without moving your eyes, reduce the number of objects in each quadrant.)

✓ ***Awareness Advice:***

With this practice comes the ability to count without counting—to see the whole without needing to add the sum of the parts. People who count quickly actually perceive larger sets, or groups of items, and therefore do not need to move their eyes slowly to count each individual item within each set or group. An example is people who work in toll booths where passengers of automobiles give them mixed denominations of coins. Toll takers often do not have time to add all the

different coins in their hand. They do not learn to count faster; instead they learn to simply see the total all at once.

3. Place seven objects in the middle of your visual field. Again, can you see them all without looking at each one separately? Can you breathe quietly enough that you can feel what you do with your eyes and in fact with your whole body to see seven at once? Now, remove two objects. Can you do it without changing your eyes? Add the two objects back in. Rest.

✓ ***Awareness Advice:***

If it is too difficult to add or subtract from seven objects without moving your eyes, reduce the number to an amount with which you are physically comfortable.

4. To your seven objects, add two more. Can you add still another? Feel your eyes. Rest.

5. Now, see only four objects in your visual field and then double that number. Can you double it again without moving your eyes? Rest.

✓ ***Awareness Advice:***

You might feel your eyes making lots of small contractions. That is normal. You are feeling what I call, the "thinking muscles," i.e., the muscles of the cilia around the iris and the ocularmotor muscles that move your eyeballs. If these muscles are inefficient and disorganized while focusing, in order to compute, we might even have to count aloud, transferring the work from our eye muscles to our mouth. This makes us even slower and less efficient. Eventually, you will feel no work at all in your eyes, except for very slight contractions you might have never felt before. But, the eyeball itself should not move.

6. Experiment with adding groups of numbers without moving your eyes. For example, can you add five plus four plus three by creating the images of the objects in your visual field and seeing them all at once? Rest. (Add together any numbers you would like in this manner.)

7. What is the largest number of objects you can hold in your visual field at once without individually counting them? Take time to let all the objects appear to you to get an answer. Do not count them individually! Breathe, stay calm, and wait until you can see your maximum number of objects at once. Can you subtract five objects from this total by letting them simply disappear? Do not count either group. Rest.

8. Experiment regularly with adding and subtracting groups of objects without moving your eyes. As you master this skill, you will find your mind able to grasp more complex calculations involving sets of numbers, as in multiplication and division. If our focusing skills and style of concentration is stuck at counting individual items and adding them together with many eye movements—and we don't learn to perceive whole patterns, groups, or sets—it becomes very difficult to perform calculations.

MOVEMENT LESSON 4

SHARPEN YOUR FOCUSING SKILLS

This lesson will give you an opportunity to feel the muscles that focus your eyes. As your eye muscles become more awake and alive they will gain an ability to focus and accommodate for depth perception and allow you to be more visually precise. You may find this lesson slightly disorienting, as most people have never had an opportunity to experience their eye muscles to this degree. Go slowly. You don't have to do the whole lesson at one time, and make sure you don't interfere with your breathing as you explore this lesson.

1. Lie on your back with your knees bent and your feet in a comfortable position. Make sure that your knees and feet are far apart. Close your eyes and see a ring of any color you choose. Very slowly, watch the ring go away from you as it lifts up toward the ceiling. Allow it to go far enough away so that you can no longer see the space inside the ring. It recedes to a tiny spot. Can you feel what your eyes are doing to activate this image?

2. Take the spot far away from you and slowly allow it to descend until it grows again into a ring. And then, let the ring descend closer and closer to your face until it grows larger. Eventually, the ring becomes like a large hoop and goes around your head. Rest. Observe what this did to your eyes.

3. Return to the ring above your eyes. Slowly send it far away until it becomes a spot again. Then, let it descend to form

a hoop around your face. Feel the work your eyes perform to create this movement. Slowly, repeat the movement of the ring back and forth far from your face, up to the ceiling and down over your head near the floor. Repeat this several times until you can feel how the work of your eye is linked to the movement of the ring.

4. Roll over onto your stomach. Let your legs be a slight distance apart. Put your forehead on the back of your hands in a place where your head can rest comfortably. Repeat the movements of the ring moving far away to become a spot and returning toward you to form the circle around your head. Do your eyes feel different doing this movement when you are on your stomach?

✓ ***Awareness Advice:***

You have an opportunity to feel the weight of your eyeballs and the work your muscles around your eyeballs perform in order to move them, as you shift from lying on your back to lying on your stomach. Go back and forth again until you can feel the difference when your eyes must lift the ring up toward your face (while on your stomach) as opposed to allowing the ring to descend over your face.

5. Come to sitting in a comfortable position. Look out horizontally and close your eyes. See your ring at some distance in front of your face and gradually move it back and forth, far from you and very near to you, until you feel not only your eyes but the corresponding movement of your head, neck, and jaw. Where in the movement is it easier to breathe? Rest.

6. Come to standing. Find an easy comfortable balance with your legs slightly apart. Again, explore the movement of the ring as it moves back and forth horizontally in front of you. Can you feel the effect of your eyes activating this image on the position of your head and perhaps even on

your balance? You might feel your entire body moving with the ring. Rest. Go for a walk and observe, as you look around, if you can still feel the work your eyes do to focus and find depth perception while they are open.

✓ ***Awareness Advice:***

It is harder to sense and feel the movement of our eyes while they are open and actively engaged. Our attention goes to what we are seeing, not to how we are using the eye muscles to see. That's why this lesson as well as the other lessons in this section should be done with the eyes closed.

SECTION VII

IMPROVING YOUR BREATHING

Breathing is one of the most certain, constant movements of our life. The rhythm of breathing, like the rhythm of our heartbeat, provides a crucial connection between the motions of our body and the emotions in our body. Most people have no awareness of whether their breathing is limited; yet in almost all activities, our breathing must constantly adapt and adjust to every movement we perform and every situation in which we are engaged. Our breathing must even adjust to our innermost thoughts and feelings. It's important not to have our breathing rigidly fixed, so we only use limited muscles of our belly, partial movement of our ribs, or find ourselves taking deep breaths without being aware of the resistance of the muscles in our torso. The necessary movements of our ribs and belly are in a constant fight with a muscular straitjacket we unconsciously impose on ourselves.

The following lessons will open you to the enormous possibilities for pleasure with each breath you take and will give you the ability to be aware of how complex a simple thing like breathing really is. When we are simply lying on our backs and taking a deep breath, over sixty muscles contract and 104 joints move a slight amount. Yet, many people breathe using only a few muscles and a few joints, so when they bend over to pick something up, they have to hold their breath because their body has forgotten how to move the muscles and bones involved in respiration as soon as they are in a slightly bent position. Many other people hold their breath in preparation for almost any action, even such simple things as getting in and out of a car or picking up a dish. In other words, with limited possibilities for the potential motions of our breathing, there is a frequent disconnection between the steadiness of our breath and our actions in the world.

MOVEMENT LESSON 1

BREATHING LESSON 1

In this lesson, we are going to learn to recognize the movements of our ribs, diaphragm, abdomen, the sides of the body, and even the back—all the parts of ourselves that move when we breathe. Because most people tend to move quite a lot in certain parts of their torso as they breathe, while other parts don't move at all, we're going to involve more parts of ourselves with the breathing motion.

1. Lie on your back, stretch out your legs, and put your arms down at your sides. Put a pillow or cushion underneath your head, or head and neck, to make yourself comfortable. Observe the movement of your breathing without changing anything. Don't inhale more or faster than usual. As soon as we begin to watch our breathing, we tend to alter it.

   What parts of your torso expand the most? What parts don't seem to be involved? If I told you I don't believe you ever breathe, how would you prove to me that you do? How do you know you're breathing? Can you identify the mechanisms of your breathing? Or do you identify with the feeling of air passing through your nostrils and going down your throat?

2. Draw in air to fill your lungs as much as you can comfortably. Breathe fully in and out this way and observe what happens to your lower back. As you inhale, do you feel any tendency for the hollow of your back to rise from the floor?

As you breathe out, does the lower back go down toward the floor? Do you feel your waistline moving out to the sides? Or does everything seem to move toward the ceiling without any tendency to reach backward or to the sides?

✓ ***Awareness Advice:***

Instead of increasing the actual volume of the chest in accordance with its structure, many people raise their chests by raising their lower back from the ground. This means not allowing the breastbone—or sternum—to move relative to the spine. As you observe yourself breathing, don't change anything. Our goal is the awareness of how we move so that we can learn to make adjustments that are more organic and far deeper than artificially imposing some idea of breathing on ourselves.

3. Bend your knees and place your feet flat on the floor, near yourself, with the feet apart, and each leg balanced. Place one hand on your belly and the other hand on your chest, keeping your elbows on the floor. Take a comfortably full breath of air and hold it. Without letting any air in or out, push that air down as though you were sending it out between your legs. Feel how your belly and waist expand. Still, without breathing, pull the belly in and let the chest expand again. Go back and forth like this, slowly pushing the air down toward the bottom of your pelvis, and then up toward the top of your chest.

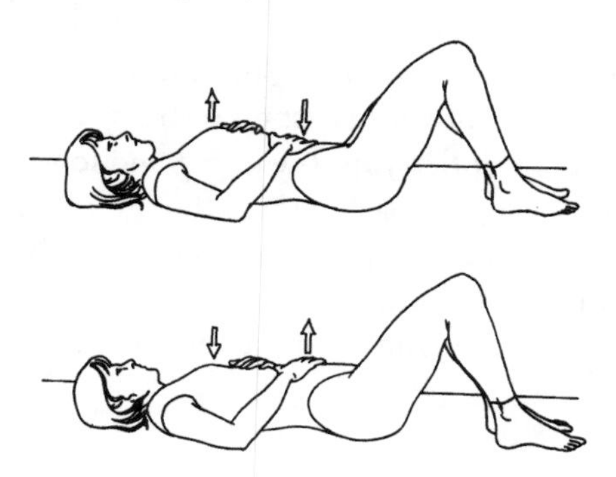

✓ ***Awareness Advice:***

Breathe whenever you need to; simply break the cycle completely and rest for awhile. Don't try to do more than five or six movements back and forth before you stop and rest.

Go back and forth slowly, gently, and steadily. You might discover that your hands go up and down, alternating like a teeter-totter, or a see-saw. Please don't challenge yourself to see how long you can hold your breath.

✓ ***Awareness Advice:***

The idea of holding your breath and pushing the air up and down is to imitate some of the movements that you make when you breathe, without actually breathing. In this way, we're going to use more muscles than most people probably use when they breathe, muscles that should be and could become involved in breathing. In a sense, you're waking up your torso to it's own potential for moving with the breath.

4. Roll onto your hands and knees and come to a crawling position. Keeping your hands where they are on the floor, lower your pelvis toward your feet, with your chest toward the floor and your legs. You can rest your head on the floor and part of your chest on or between your thighs. In this position, notice your breathing and lower back. It may be that now your back feels as though it's lifting to the ceiling and out to the sides. As you take deep breaths here, you might even feel your back expanding all the way down to your hips. Leave it and rest on your back.

5. Once again, bend your knees and balance them above your feet. Put your hands on your waistline so your thumb is toward your back and your fingers are around your waist toward the front. Breathe in again and hold it. Pass the air down to your pelvis. Does it feel like you are pushing out in all directions with your belly, not just to the ceiling?

   Can you feel the distance between your fingers and thumbs widen as you push? Now pull your belly in to expand your chest. Can you feel your thumbs and fingers coming together? Pass the air back and forth through your torso several times while observing how your back actually reaches to the floor, as well as going forward and to the sides. Remember, this is not a challenge to see how long you can hold your breath.

✓ ***Awareness Advice:***

End this lesson by breathing normally for a while. It's all right if "normally" might seem rather confusing right now. Your breathing will be altered by this lesson. Observe how it's different from before.

MOVEMENT LESSON 2

Breathing Lesson 2

This lesson will expand upon some of the themes you began in Breathing Lesson 1. Be sure you have completed Breathing Lesson 1 before doing this lesson.

1. As you breathe in, expand only your chest while pulling your belly toward your spine. Exhale by collapsing your chest and pushing your belly out in all directions. This is going to be the way you will breathe awhile. Make sure you breathe with a normal volume of air and at a normal speed. Try to make this seem as though it were normal.

✓ ***Awareness Advice:***

For some people this way of breathing is normal. Also, in certain situations, this breathing is normal for everyone. For example, if we need to push against something forcefully or move very suddenly, we'll tend to push our belly out when we exhale, while inhaling in the chest. Babies often breathe this way normally. Try to establish a feeling of ease as you do this so that there's no strain.

2. The next time you breathe in, hold your breath. Notice the expansion in the ribs between your arms and the sides of the ribcage. Push the air down toward your pelvis and up to your chest as in Breathing Lesson 1, but this time do it quickly.

✓ *Awareness Advice:*

Make sure not to take in as much air as possible so this movement can be comfortable. Feel the liquid contents of your abdomen pushing out in all directions like a hydraulic system. See that your throat doesn't have to strain. Pause and breathe whenever you want to, and then return to this movement. Teach yourself to make it light.

Leave it and just breathe normally for awhile.

3. Roll over and lie on your stomach, legs easily apart, hands on the floor near your head. Put your head in any position that's comfortable. Feel the front surface of your body against the floor. Inhale and hold your breath, and then alternately push the air again from chest to pelvis as you did on your back. Do you feel symmetrical or do the muscles in your belly seem to have more strength and movement on one side? Pause.

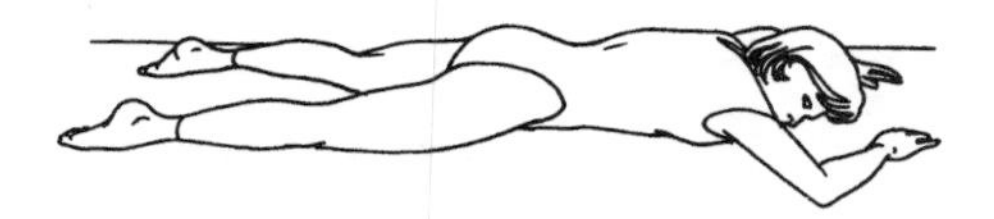

4. Breathe in and expand only your chest; exhale and push your belly out again. But, only think of breathing in the right side of your chest, as though you are going to expand the ribs independently in the right side. When you exhale, think of pushing out only the left side of your belly. You'll find your body will rock diagonally. Rest.

✓ *Awareness Advice:*

Remember to do this at normal speed and normal volume, but make the pushing movements against the floor distinct.

5. Turn your head to the other side. Now breathe in on the left side of your chest and breathe out pushing the right side of your belly against the floor. Imagine the air passing on a diagonal from chest to belly. Is this an easier diagonal than the other?

6. Inhale and hold your breath comfortably. Push the air back and forth between your chest and belly and notice how much your back rises to the ceiling when you push out your belly. Notice if you can detect whether your breathing is symmetrical or if you still feel stronger on one side of the ribcage or one side of the belly. Roll over and rest on your back.

7. Bend your right knee so you can stand on your right foot. Bend to the right so your right hand can reach down toward your right foot. Rest your left arm on the floor above your head. Let your head turn to look toward your left arm. Inhale in your chest, pulling your belly in. Then, exhale while pushing out your belly. Make this breathing easy and comfortable without exaggerating the volume of air. Pause for a moment.

8. Inhale and hold your breath. Push the air back and forth from chest to belly. Rest.

9. Blow out your air so you have only residual air in your lungs, and then hold your breath. Once again, pass this small volume of air up and down from chest to pelvis.

   Take a full rest with your arms and legs down. Make yourself somewhat symmetrical and observe whether something has been awakened on one side of your body that's not yet awake on the other side.

10. Draw your left knee up and stand on the left foot. Rest your right arm on the floor above your head, turning your face toward the arm, and repeat steps 7-9 slowly and carefully on this side.

Take a few moments to observe changes between the two sides of your body and to enjoy the feeling of expansion.

MOVEMENT LESSON 3

BREATHING LESSON 3

In this lesson, you will learn to expand your torso evenly, in all directions, when you breathe. You will also learn how to improve your balance and your posture by letting your body expand evenly when you breathe, rather than only forward and backward. Many people find this lesson the most significant of all the lessons in this book.

However, make sure you fully learn Breathing Lessons 1 and 2 before proceeding.

1. Begin by lying on your back. While you're resting, breathe slowly in small steps so you make many movements of your chest and abdomen that are distinct, rather than taking all the air in or out in one gulp. Does this make you more aware of what you use to breathe? Try breathing this way for several minutes. How many steps can you make both inhaling and exhaling? Rest and breathe normally.

2. Roll over on your hands and knees in a crawling position. Is your back level, arched up to the ceiling, or curved down to the floor? Inhale in your chest, hold your breath, and pass the air back and forth from chest to pelvis. What happens to your pelvis? Does the position of your body alter as you pass the air back and forth?

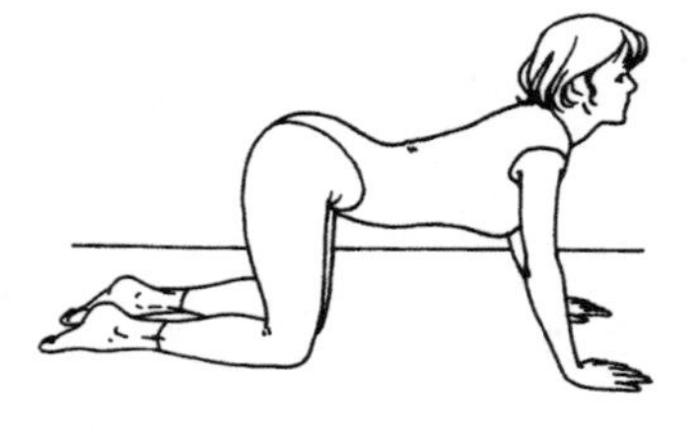

3. Keep your hands on the floor and lower your pelvis toward your heels while lowering your chest to your knees and your head to the floor. Rest this way. Now inhale in your chest and when you exhale, push out your belly and feel how the back moves. Can you feel how the sides of the body push out?

4. Hold your breath with a comfortable lung full of air and gently pass the air back and forth from upper back to lower back. Don't strain or hold your breath too long and don't go too quickly.

✓ ***Awareness Advice:***

Make sure you're moving at a speed that allows you to feel very fully how the back moves, as though it were doing all the breathing for you. Rest on your back.

5. Come onto your hands and knees in a crawling position. Let your belly slump down to the floor so your back is bowed. Inhale, hold your breath, and push the air from chest to belly. Is that efficient here or quite difficult?

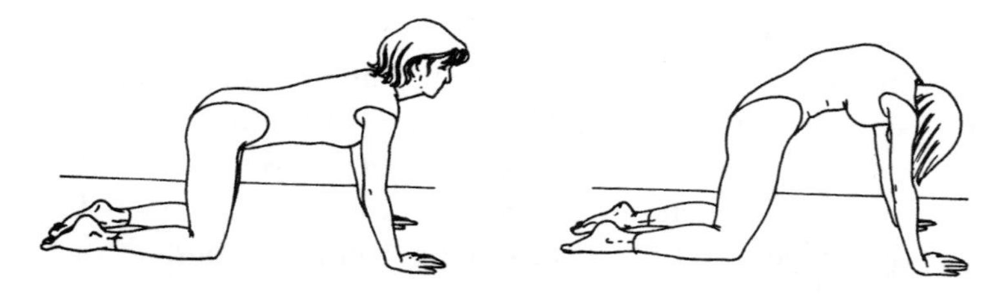

Now arch your back up to the ceiling. Inhale, hold your breath, and again push the air back and forth from chest to belly. Is this difficult or easy? Rest in any position that's comfortable.

6. Return to a crawling position with your elbows straight and your knees a comfortable distance apart. Place your spine at a level where you feel very even and long across the back, and where you are neither bowed nor arched. Inhale again, hold

your breath, and pass the air from chest to belly. Experience your body as a cylinder, moving evenly in all directions.

✓ ***Awareness Advice:***

Once you've found that place where your back supports you and your breathing helps support your back, then simply breathe in your chest and out your belly so you feel the expansion in your upper and lower back as much as in your chest and belly. If this is hard on your wrist, rest and then continue. Make sure you learn this well: how posture can be impro[illegible] breathing and how breathing can suppo[illegible]

Leave it [illegible] ow much more

[illegible] you c[illegible] and, once again, pass t[illegible] up to your chest. Can you fe[illegible] ing away from you and the right side of y[illegible] t off from the floor as you push down into the pelvis? Rest on your back.

✓ ***Awareness Advice:***

Feel how you can rock your hip away from you and toward yourself by passing the air up and down through your body. If the pelvis doesn't move very much, then move it intentionally. You could place your other hand against your hip to help feel it go down and up.

8. Repeat step 7 on the left side, leaning on your left elbow. Rest on your back.

9. Come to sitting with your legs crossed. Give yourself a big hug by holding your hands over your shoulders with crossed arms. Drop your head and let everything hang. Take a deep breath of air and feel the expansion in your back. Notice that every time you breathe in, you'll feel a tendency for the body to rise, as though your breathing is pushing your back up. Rest again.

10. Come to standing and give yourself a moment to settle on your feet. Explore how your back moves with your breathing while standing.

✓ ***Awareness Advice:***

In ideal breathing, we expand evenly in all directions, like a cylinder—not just forward and backward in our belly and chest, which throws us off balance every time we breathe in and out. Can you experience your back and your breathing mutually cooperating to support your posture? Walk around that way and play with it every day of your life.

SECTION VIII

The Fast Road to Stress Relief

When under stress many people seek methods of relief such as: alcohol, drugs, compulsive behaviors, or whatever seems to work to relieve the tensions that accumulate in the course of our days. In this chapter you will find many lessons that can quickly provide not only the feeling of relief from stress, but for many, a sense of well-being and pleasure. We rarely think of using unusual movements of our body as a major way to obtain highly relaxing states of mind. However, our body not only accumulates stress, it can also accumulate pleasure. Some of the lessons focus on relaxing tight muscle groups, some focus on using the body against the floor as if you were getting a good massage; and others focus on alleviating the muscular tensions that occur around our joints that make the feelings of stress so difficult and even painful. Some people use the lessons in this chapter on a daily basis as soon as they come home from work or in the middle of the day—perhaps after a stressful meeting. Others use them before they go to bed at night, to relax as much as possible before sleep, and others use them as their good-morning-wake-up exercise, to get ready for a hard day. In any case, you can use all of them any time you want, and feel free to invent new movements as you explore your body with these lessons. There are no side effects but pleasure.

MOVEMENT LESSON 1

RELEASING THE HIPS INTO PLEASURE

Many people find that the inside of their legs, as well as other muscles around the hip joint, are stiff. In this lesson, you'll find a comfortable and pleasurable way of releasing excess work in these muscles.

1. Lie on your back with your arms down and your legs straight. Feel the pressure under your heels. Does the pressure on your left heel differ from your right? Is the pressure more on the outside of one heel than the other? Also, observe the pressure on the calf muscles. Does it tell you which of your two legs is pointed more to the ceiling and which is pointed more to the outside? Make sure your legs are relaxed. Don't try to hold them in any particular way.

2. Very gently, turn your right leg further to the right and let the knee softly bend, so that the outside edge of your right foot begins sliding on the floor up toward you. Slide the edge of the foot back down the same track it went up. Repeat the movement several times, very slowly, searching for the path of least resistance and effort as you slide the outside edge of the foot up and down on the floor.

✓ ***Awareness Advice:***

As you slide the foot up and down, make sure the knee hangs open completely, so there's no work in the inner muscles of your thigh. Every so often, let the leg rest with the knee bent and the foot pulled up to make sure you let go of the inner thigh muscles.

Many people overwork the muscles on the inside and outside of their thighs using far too many muscles that do not support the simple action of bending the hip and knee or straightening the hip and knee. The more unnecessary muscular work you perform, the more difficult, heavy, and resistant the movement of the leg becomes. So be sure to concentrate on releasing the inner muscles of your leg as you perform this movement.

After exploring sliding the edge of your foot up and down the floor, rest and observe what's changed in the way the leg is resting on the ground. Is the leg softer? Is it pointing in a different direction than earlier? Does the hip feel softer? You might even find that your lower back has released to the floor on the right side.

3. Perform the same exploration of this soft and lazy movement on the left side. Is it easier or harder than the right? Notice if your eyes, or even your head, pull to the left side when you move the leg up and down. Rest and feel the expansion in your pelvis and the release of excess work in your legs.

4. Turn both of your legs open, with the knees hanging apart, and slide the outside edges of both of your feet up toward yourself at the same time. As you go up and down this way, slowly and gently, feel what your back and your pelvis do to assist the movement.

✓ ***Awareness Advice:***

As the edges of the feet slide up, try to keep the space between your feet about a foot apart.

5. Rest with your arms on the floor above your head and out to your side, with the elbows slightly bent. Once again, slide the outside edges of both feet up toward yourself and leave them there with your knees hanging open. Now, let your head easily roll from side to side while breathing deeply.

   Rest in this position, with the knees suspended and the arms open, and feel the comfort of being like a baby sleeping on the floor. Let your ribs move freely as you breathe.

MOVEMENT LESSON 2

ROLLING INTO RELAXATION

Rolling is one of the most wonderful ways we have to refresh and relax our bodies. Not only does it improve our coordination by integrating our arms and legs with our torso, but as we roll across the floor, we're also getting a free massage as the floor presses on our muscles.

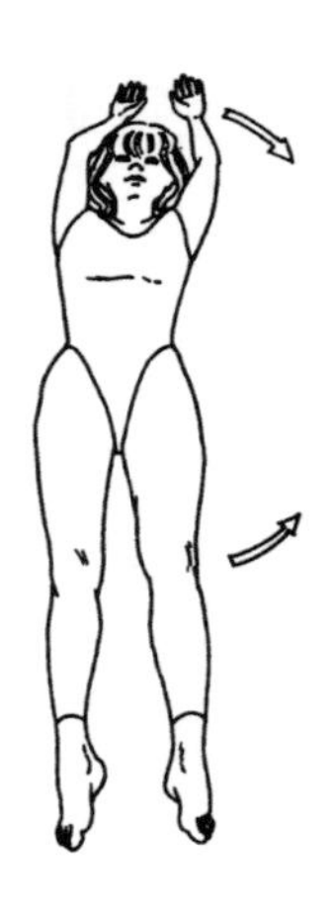

1. Rest with your legs straight and your arms on the floor above your head. Take a moment to observe the quality of softness in your body against the ground.
2. Slide the back of both of your hands on the floor to the left while sliding both of your feet to the left at the same time. Allow your head and pelvis to turn to the left at the same time as well. Can you rest on your left side for a moment?
3. Keeping your feet and hands heavy on the floor, sweep your arms in a long lazy arc back over your head as you slide your legs long, turning yourself onto your back.

✓ ***Awareness Advice:***

As you sweep your arms and legs to the left, feel that all you need to do is turn your legs at the hips so that the outside edge of the left foot and the inner edge of the right foot can slide on the floor. Feel the pressure of your body against the ground moving constantly over to the left side.

As you move back, be sure to sweep the hands in a long reaching arc above your head, as your legs slide straight at the same time. Rest with your arms down at your sides and feel the difference between the two sides of your body.

4. Sweep your arms and legs to the right and roll to the right side in the same way you did to the left. Which side feels more coordinated? Rest with your arms down.

5. Rest your arms on the floor above the head again and now sweep your arms and legs from side to side as you roll from right to left.

✓ ***Awareness Advice:***

If you feel difficulty breathing, or you feel stiff while doing this lesson, put all of your awareness into softening your body so that there's no sharp effort in any particular place.

Rest and enjoy the pleasure in your body and the improved quality of your contact with the floor. Get up very slowly to stand and walk. It may take some time to focus on the day's activities.

MOVEMENT LESSON 3

Rolling into Length across Your Stomach

This is a delightful extension of the movement ideas in Movement Lesson 2 "Rolling into Relaxation." This lesson will be done best after completing that one.

1. Lie on your stomach with your legs slightly apart and your arms on the floor above your head. Lift your left hip from the floor and slowly roll toward your right hip. As you do this, you'll experience your left knee and left elbow bending slightly as the left hand and foot move toward each other. Do this movement several times and then rest.

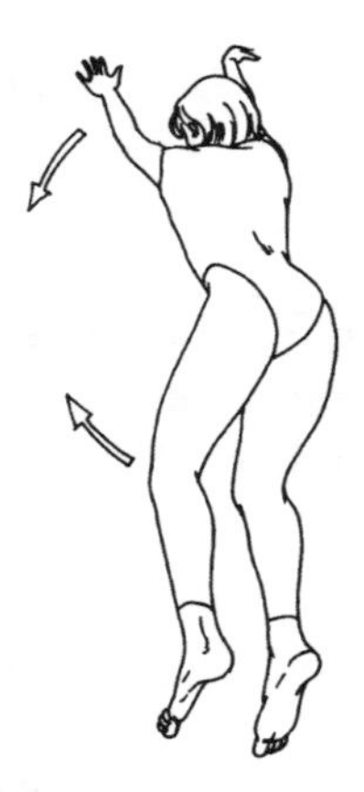

2. The next time you lift your left hip, slide your right arm on the floor underneath your head to roll yourself further onto your right side. As you slide your right knee and foot up

underneath your left leg, you will find yourself lying on your right side with hips and knees softly bent and both arms in front of you.

3. Keeping your hands and feet separate from each other, slide them away from each other on the floor as you sweep your right arm underneath your head to roll back onto your stomach. Make sure you arrive with arms and legs long.

4. Practice the coordination and timing of lifting the left hip, sliding the right arm under your head in a large arc, and drawing your legs up to roll you onto your right side several times. As you roll back again onto your stomach, feel the arms, legs, and trunk all moving at once.

✓ ***Awareness Advice:***

If the movement feels effortful, as if the floor provided too much friction, play with integrating the actions of your pelvis, arms and hands, and legs and feet, together continuously. You may have to emphasize some parts of your body at a slightly different time than is spontaneous at first. Eventually, the movement should feel like a very light and easy way to roll onto your side and return to your stomach, but it does require deepening your sensations and using your fullest attention and awareness.

5. Perform the same action rolling from your stomach to the left side and back. Which side is easier? Rest on your back. Observe your breathing and the comfort throughout your body.

6. Return to your stomach, as before, and sweep your arms from side to side as you roll from one side to the other across your stomach. Seek pleasure as you move. Roll over, rest on your back and enjoy the sensation.

MOVEMENT LESSON 4

LENGTHENING HAMSTRINGS, I

The hamstrings are important because they flex the knee; when walking they pull the foot backward at the knee, while they also contribute to extending the hips slightly along with the gluteal muscles of the buttocks. They contribute to stabilizing our knees when we push with our feet, as in jumping. Because these are very common activities, these muscles tend to get short and tight. Unfortunately, most exercise systems simply have us pull against the muscle to stretch it until it gives up. In this lesson you'll learn how to comfortably and enjoyably lengthen your hamstrings so the muscle learns to retain its length.

1. Lie on your back with your legs long and feel the distance of the backs of your legs from the floor. The tighter your hamstrings the higher your legs will be. Your two legs might feel different from each other.

2. Roll onto one side and sit up. Sit with your left leg in back of you and your right leg straight out in front. Put your left foot in a comfortable place in back with the left knee slightly forward. If this is uncomfortable, bring your left foot near your crotch in front of you, with the left knee hanging open toward the floor. You can lean on your right hand for support.

3. Put your left hand on the outside of your right thigh and slowly massage your right leg by kneading and rubbing the muscles, slipping over them with your hands. Massage in long, slow strokes down toward your ankle.

✓ ***Awareness Advice:***

Feel the texture of your leg, as well as its shape, as though you were exploring it for the first time. Make long, steady strokes down toward the ankle and back toward the hip. Pay no attention to how far you're going. Notice that if you let your shoulder release, and allow your body to turn to the right, it's quite easy to go further.

4. Put the right leg out in front of you again, with a straight knee. Put the leg at a different angle to your trunk. Begin massaging the inside of the right leg with your left hand and experience how you rock back and forth on your pelvis as you reach with your left shoulder and your head. If you happen to touch your foot, fine. If you can only go just below your knee, that's also fine. Let your knee bend whenever it wants to so the right leg is soft and can adjust easily to your massaging. See if you can massage some part of your foot, and perhaps play with your toes. Allow the knee to bend to help you. Rest on your back and feel the difference in your two legs.

5. Roll over and sit again. This time put the right foot in back of yourself, and the left leg straight in front of you, so you're sitting as a mirror image of before. Use your left hand for support on the floor and begin massaging your left leg thoroughly with your right hand. Rest on your back again and feel the change.

✓ ***Awareness Advice:***

You'll have more weight on your left hip now. Remember that if the position is difficult, you can sit with the right foot near your crotch, with the right knee hanging open. Let your head and neck relax completely as you rock back and forth toward and away from your foot. Every so often, play with your toes, allowing the knee to bend so you can do so. You may never before in your life have taken the time to slowly and carefully massage your legs. Often we know nothing about our bodies until something hurts and goes wrong. Only then do we look at ourselves carefully and worry about it.

6. Come to sitting again, this time leaning on your hands comfortably in back of you, with your legs straight in front of you, toss your legs wide apart, and bring them back again several times. It should be like opening and closing scissors. Now leave them wide apart wherever they easily go. Put both hands on your right leg near the hip, and begin to massage down the leg toward the foot in long, languid strokes. Let yourself breathe out as you go down the leg.

7. The next time your hands return to your hips let them slide over your belly as well as your back to the left hip, and begin to massage down the left leg and back again. Think of yourself as a sculptor, shaping your legs as you massage. As you go down one leg and then the other, think of passing your hands across your body, as if you were sculpting the connection of your legs to your torso. Go in a smooth and steady motion.

✓ ***Awareness Advice:***

Let your head, neck, and shoulders be soft and relaxed as you massage. Let your pelvis rock from side to side on the floor. Every so often, vary what you do with your feet by pulling your feet back and keeping your knees straight. Remember your goal is to let the movement be sensual, as if you were making a sculpture of your lower body. Let the knees be soft and bend as they need to.

Rest on your back and feel how close your legs and back are to the floor.

MOVEMENT LESSON 5

LENGTHENING HAMSTRINGS, II

In this lesson, you'll learn another way to lengthen the backs of your legs as well as all the muscles that extend your back and hips.

1. Lie on your back for a few moments and feel your contact with the floor. Which side of your lower back is higher from the ground? Which is higher, the backs of your knees or your lower back? Evaluate this without moving your hands to measure.

2. Sit up with the bottoms of your feet on the floor and your knees bent, close to your chest. With your right hand, hold on to the outside edge of your right foot. Use your thumb and fingers together to get a firm grip. Put your other hand wherever is comfortable for support; you might want to hold on to your left knee. Find a way to be comfortable in this position. Make sure your right knee is inside of your right arm so the right arm is not between your legs but around the outside of your right leg.

3. Put any part of your face on your right knee, and rub your face against your knee as if you were washing it with your knee. Touch your chin, both cheeks, and your forehead on it.

✓ ***Awareness Advice:***

If the position is not comfortable for you, let your left knee hang open and bent to the side.

4. Keeping your forehead on your knee, and holding on to your right foot, slide the right heel away from yourself until your knee goes straighter. Only go as far as you can before your forehead wants to separate from your knee. Come back up again and go back and forth this way several times. Rest on your back for a moment.

✓ *Awareness Advice:*

See how long your head can stay attached to your knee. If it wants to separate at a certain point, where is that? Feel your pelvis rocking forward and backward as you slide the foot back and forth.

5. Sit again the same way. Hold on to the outside of the right foot with the right hand again and put your chin on your right knee. Once again, slide the foot away from yourself and back again. Can you feel the difference in your back when you have your chin against your knee, as opposed to your forehead? Pause. Now, put one of your cheeks against your knee as you continue sliding your foot. Does one cheek make it easier to do the movement than the other?

✓ *Awareness Advice:*

Can you feel how you use your back to slide the foot back again? Know where your face and knee separate as you lengthen the leg, and also know what to do in order to pull the foot back.

Rest on your back and feel the difference between the backs of your knees, hamstrings, and lower back. For example, you might find the right side of your lower back is closer to the floor than the left side. Can you notice the difference?

6. Bring your weight up to sitting again and hold on to the outside edge of your left foot with your left hand. Read instruction 2 carefully, as you will be mirroring that exact position. Repeat instructions 2-5 on the left side. Which side is easier? Rest on your back again and observe if it has gotten flatter.

7. Sit again with your feet on the floor and your knees near your chest. Hold on with your right hand to the outside edge of your right foot and with your left hand to the outside edge of your left foot. Make sure your knees are in between your two arms. Move your head toward both knees. There's a groove formed between both knees that's shaped perfectly for your forehead or your nose. If you can't get your head there, that's fine. Just let the head hang.

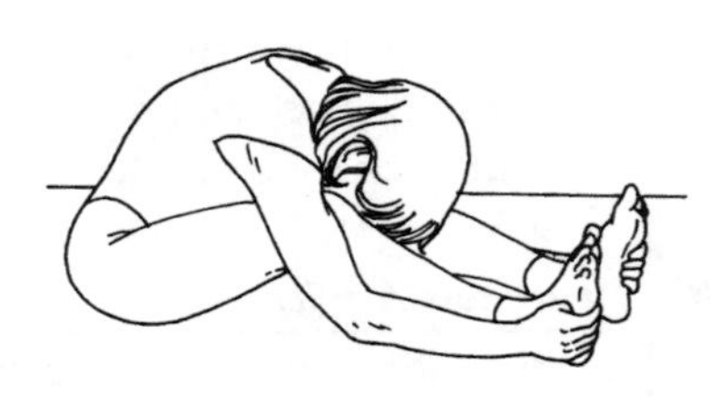

   Push both feet away from yourself sliding both heels on the floor. Let your head stay in contact with your knees as long as possible and come back again.

   Rest on your back and feel if your spine and legs are closer to the floor.

✓ ***Awareness Advice:***

This is a challenge to your back, so take your time as you go back and forth, and go only as far as is comfortable to pull back again easily. Feel that your breathing can be easy and steady throughout the movement.

✓ ***Awareness Advice:***

If you get a slight aching feeling in your back, all you have to do is draw your knees above your chest for a moment and you'll find the aching will disappear immediately. You can also review any part of Movement Lessons 1 or 3 to relieve any distress. If you have a slight aching feeling, it stems from the muscles being much longer than is familiar to you. The sensation goes away once you get used to doing this lesson and the muscle tone learns to readjust with more length.

8. Sit again and now reach both of your hands between your knees to hold on to the outside edges of your feet. Let the knees hang open with the feet near your body. Breathe comfortably and relax your shoulders.

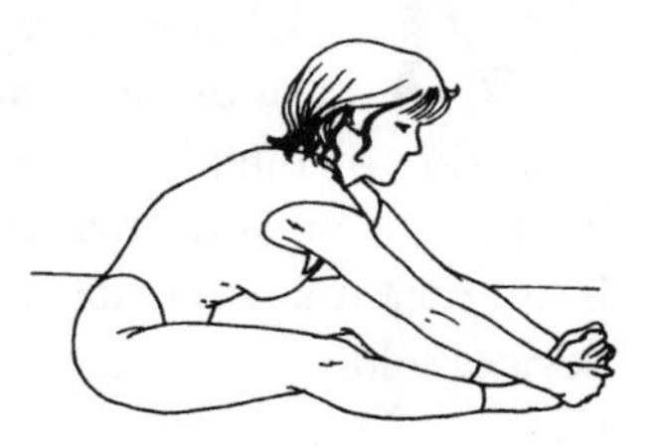

While holding onto the edges of your feet, slide them away from you until the legs become straight, and then bring them back again. Rest on your back. When you stand and walk, feel what difference there is for you.

MOVEMENT LESSON 6

MASSAGE TO GREATER LENGTH

Many people get some form of massage for stress relief, however, that's not always convenient. Most of us do not realize we can create the same depth of relaxation by ourselves. In this lesson, you'll learn how to use your hands on your own body, while doing movements. This is not meant as a replacement for all the various mechanical devices sold to knead and vibrate muscles, but you might find it more effective.

1. Lie on your back with your arms and legs long. Close your eyes to help you notice what parts of your body feel like they are pressing against the floor. Your pressure pattern is like your signature; it is unique to you. Observe how this "pressure signature" changes during the lesson.

2. Bend your left knee and put your foot on the floor. Gently push down through your foot so you roll your weight to the right hip and lift the left side of the pelvis. Keep your left knee vertical. Notice if you feel any changes in your left ribs and arm. The Feldenkrais Method is a continuous process of exploration and discovery. Notice where is the best place to put your left foot in order to push comfortably. Allow the right leg to relax as the left leg pushes. Rest with your arms at your sides and your legs long. Notice if there is a change in the pressure pattern.

3. Rest your arms on the floor above your head, but wide apart, and open your legs so that you are lying on the floor, forming a large "X" with your body. Bring your left knee up and place your foot flat on the floor. Take your left hand and begin massaging the right shoulder and continue down the arm toward the hand. Push through the left foot to help you reach the right hand. Let the massage move up and down from your chest, to your shoulder, elbow, hand and thumb. Go easily back and forth a few times in long, slow strokes. Rest, stretch out flat and notice any changes in your pressure pattern on the floor.

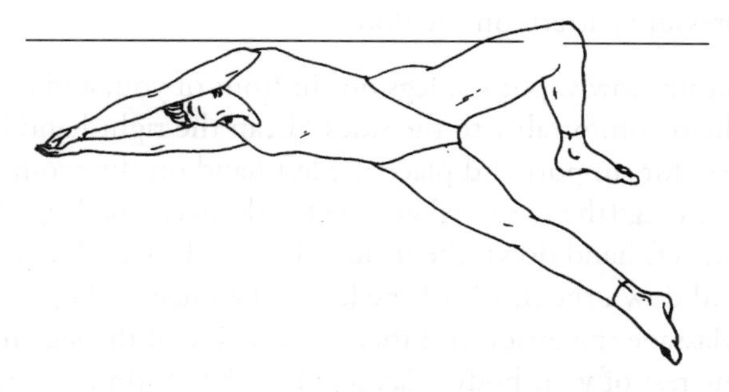

4. Put your arms out to the side and a little above your head with the back of the hands on the floor. Bend your right knee and stand the right foot on the floor. Push through the right foot so your weight rolls to the left side. Keep the right knee still as you push. Explore where to place the foot to get the best push. Roll with gentle effortlessness back and forth allowing the head to follow with the motion. Rest with the arms and legs long and notice any changes in your pressure pattern on the floor.

5. Put your arms out to the side again with the backs of the hands on the floor so you are lying in a large "X". Bend the right knee and stand the right foot on the floor. Put the right hand on the left shoulder. Gently massage the shoulder, up and down the arm. If you are ready to go farther down your body then push with your right foot and let your head follow the motion. Make long, smooth strokes up and down your arm, letting the right hip lift to follow the motion. Make sure that the left side of the body is still and completely at rest. As you finish massaging the arm don't forget to include the thumb. Rest with your arms and legs down and notice any changes in your pressure pattern on the floor.

6. Sit up slowly, put the legs out in front of you, and open them comfortably to the sides. Lean the right hand behind you for support and place the left hand on the groin of the right leg (the crease where the trunk meets the leg). Slide the left hand down the inside of the right leg. Massage up and down the inside of the leg to the knee, or beyond. Massage the inside and then the outside of the leg and let the rest of your body relax to allow the motion. To make the massaging down the leg easier you can place some towels under the pelvis to sit a little higher off the floor. Allow the right knee to bend a little so you can massage down the leg to the foot, and include the toes if that is comfortable. Try massaging from the hip all the way down the right leg as far as is easy. Rest on your back and note any changes in how you rest on the floor.

7. Come to sitting again, and repeat the experiment of massaging in the sitting position, and this time massage the left leg with the right hand.

8. Sitting, bring your right palm to your belly and the left hand on your back. Take both hands and massage all the way down and up the left leg. As the hands come back to the starting position, switch so the left palm goes on the belly and the right hand is on the back. Take both hands and massage down and up the right leg. Alternate the massage by switching the hand position as described and repeat several times. Roll to your side and rest on your back. Rest with your arms and legs down noticing how you rest on the floor. Compare your pressure signature now with what it was at the beginning of the lesson. Take all the time you need to rest before you get up.

MOVEMENT LESSON 7

EXPANDING, REACHING, AND TURNING

When stress is present, people tend to tighten in the middle of the body. The reflex tightness is supposed to be only a temporary protection during a crisis. What happens most often, however, is that the tightness remains and becomes a habit. In this lesson you will explore different ways to increase the ease of motion in your waist. The movement of the arm or the head requires participation throughout the body if strain is to be avoided. Reaching more easily involves understanding the connection of the arms to the torso and pelvis.

1. Lie on your left side with your hips and knees bent at right angles and your arms at chest level straight out in front of you. See that the palms and fingers of your hands are together and resting, and make sure that your elbows are straight.

✓ ***Awareness Advice:***

If your neck is under strain while lying on your side, put a folded towel under you so that your neck is supported. If it does not seem possible for both elbows to rest straight, you might need to adjust your side-lying position. If you are rolled slightly forward, the top elbow will have to bend. If you roll backward too far the fingertips won't reach each other easily.

Slowly slide your right palm along your left palm reaching for the carpet in front of you. Repeat this motion several times, sliding your right palm forward, then back along your upper arm. Feel the texture of your arm and the carpet. You will notice that your range of motion expands naturally as you repeat the movement. Where does the ability to reach further come from? Does your back assist? Does your pelvis?

2. Next time you slide your right palm along your left, bring the right hand back along your body all the way across your chest. Make sure you are performing long, lazy movements. Do not move any further than you can do while still breathing comfortably. It is not worth it for you to compromise easy, natural breathing. Repeat this movement several times, then leave it and rest long and comfortably on your back.

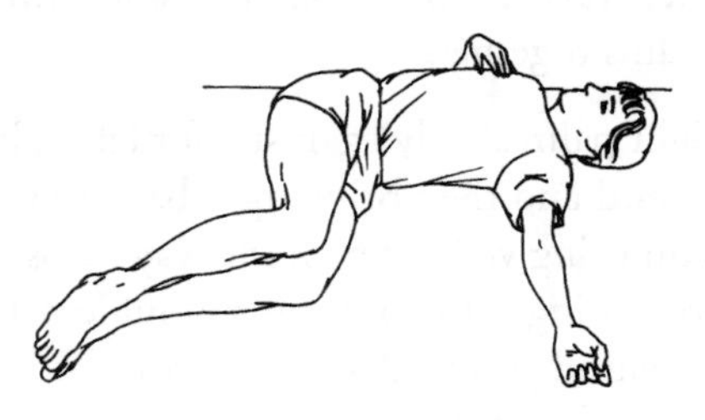

✓ ***Awareness Advice:***

Make sure your hand is always touching some part of your body softly and fully.

3. This time, roll on your side in the same position, knees and feet touching, bringing your right hand to cover your forehead, with your elbow pointing straight in the air. Roll your head slowly toward the floor and back feeling your elbow move through space. What happens to your waist? What happens in your ribs? The next time you roll, see if you can gently bring your elbow all the way to touch the floor in front of you. When you roll your head back, can you bring your elbow behind you? Feel the opening in

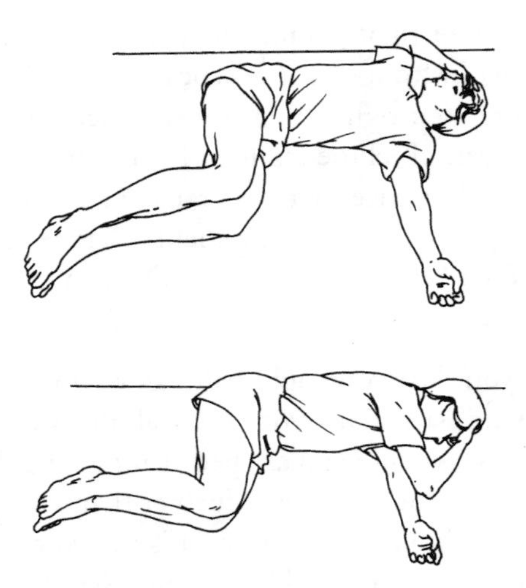

your chest and the movement in your waist. Don't strain to do this motion. Instead, move more slowly and carefully each time. Take a full rest on your side with your elbows and palms together.

4. Glide your right palm slowly against your left palm, moving it forward and then bringing it back toward your chest. Can you bring your arm all the way across your chest? Can you bring it to your right shoulder? Can you bring it back behind you so that the back of your right hand rests on the floor? To do this, allow your knees to separate if they want to, and feel the movement of your pelvis against the floor. Take a full rest on your back.

5. Roll onto your right side with your hips and knees at right angles, your arms level with your chest and bent at the elbows with your palms and fingertips together. Slide your left palm slowly along the right onto the carpet, then bring it back along your upper arm. Can you bring your left arm across your chest? Is it easier on this side? Can you bring it behind you? Rest with your fingertips touching.

6. Bring your left hand to your forehead with your palm flat against your forehead and your elbow pointing straight in the air. Roll your head a little bit forward, then back. Feel

the direction of your elbow in the air. Can you feel the inside of your mouth? The back of your right knee? Roll gently forward and rest on your side with your left palm against your forehead and your elbow touching the floor.

7. Roll onto your left side. Organize yourself in the same angular manner. This time, clasp your hands together with your fingers interlocking. Gently slide your right knee forward and backward over the left knee, keeping your feet together. Let your pelvis and waistline turn while you keep your elbows straight. Can you feel your pelvis clearly rolling? Rest on your side with your palms flat against each other.

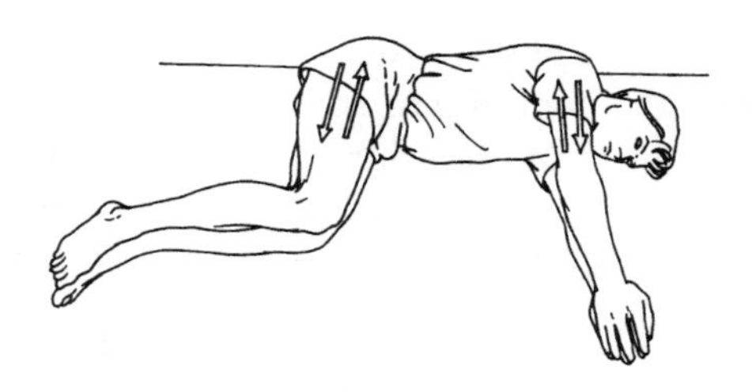

8. Now slide your right hand against the left toward the carpet while sliding the right knee back against the left. Then, as you bring the right hand back, roll the right knee forward. Repeat this motion several times until it becomes easier to do this opposition of arm and leg. Make sure you can breathe comfortably. Give yourself time to learn. Rest long and comfortably on your back.

9. Roll onto your right side with your elbows straight in front of you, hands interlocked. Slide your left knee forward and back against the right knee, using your pelvis and your back to help. In your own time, repeat these opposing movements of arms and legs. Do not strain, let it happen gently. Take a long, full rest on your back.

MOVEMENT LESSON 8

Oiling Your Hips the Easy Way

In this lesson, you'll learn an enjoyable and exhilarating way to release stiffness in your hips, knees, and lower back.

1. Take a moment to observe your contact with the floor. Calm down and focus your attention inside your body. Then, come to sitting with your knees bent so the bottoms of your feet stand on the floor. Take your left hand and reach around the outside of your left knee holding on to the outside edge of your foot, near your toes, thumb together with your other fingers. Lift your foot in the air and set it down a few times and feel what your pelvis and lower back need to do to help you. Feel free to use your right hand on the floor for support.

   Each time you lift the foot up and set it down, have the knee cross underneath your arm so it flips to the outside of the arm and back to the inside several times. Do it slowly so you can feel what happens in your hip joint to create this turning of the knee under the arm. Rest on your back.

2. Come to sitting and hold on to the foot arranged exactly the same way as before. Lift the foot up and set it down somewhere else on the floor. In fact, lift it up and set it down in as many different places on the floor as you can.

Cross it to the other side of your body and even find a couple of ways to set the foot behind you, over your head. Rest on your back again.

3. Now, arrange yourself the same way with your right hand reaching around the outside of your right knee and holding on to the outside edge of your foot. As in step 1, explore how to flip the knee back and forth under the right elbow. Rest.

4. Sit again with the right hand holding the outside edge of the right foot. As in step 2, explore all the places you can touch on the floor around your body with this foot. Rest on your back with your legs long.

5. Come to sitting with both of your knees pointing to your right and both of your feet to the left. This means you will be side sitting with your left foot just in back of you on the floor and the right foot relating to the left knee. In fact, the left knee will probably be touching the bottom of the right foot.

At the same time, hold on to the outside edge of your left foot with your left hand and the outside edge of your right foot with your right hand. Lift both feet from the floor and simultaneously flip both of your knees under both of your elbows. As the knees get straighter, come to side sitting in a mirror image of your starting position. Your knees are now to the left and your feet to the right, with your left foot touching your right thigh.

✓ ***Awareness Advice:***

As you move your feet from one side of your body to the other, in side sitting, you will experience yourself rolling on your pelvis from side to side and you will need to lift your feet high in the air for the knees to flip under the elbows. Make sure you don't strain or hold your breath. Work to make it easy.

Take your feet and knees from side to side this way, several times. Make it easier each time. Rest on your back with your arms and legs long.

6. Come to side sitting again. Choose your favorite side. Again, hold on with both hands to both outside edges of your feet. Once again, you'll be changing to side sitting on the other side by flipping your knees under your arms, but this time try to do it by sliding your feet on the floor the entire time, as the knees change from side to side. Rest on your back.

✓ ***Awareness Advice:***

As you do this movement, you will discover that your feet slide out away from you and back again along the same line on the floor while the knees flip from side to side and your pelvis rolls from side to side.

When you get up to walk, notice if your legs and back have more freedom.

Awareness Advice (for the Long Lesson Ahead)

By now you will have experienced some of the benefits of the Feldenkrais Method. You will have felt how long-standing limitations can melt away once you know *how* to use yourself in other, more effective and gentle ways. This first-hand experience can give you what no amount of testimonials, advice, or assurance can: an idea of what is possible once you start using the intelligence of your body.

Don't limit this knowledge to when you are doing the lessons. Most of us don't have time to spend an hour each day to stay in shape . . . nor do we need to. Begin to make all those mundane, everyday activities that you perform anyway throughout your waking hours into your practice. Every time you walk, drive, sit down, get up, bend, kneel, reach, twist, or turn, choose to do it with awareness, using your body's intelligence. Think of the various distinctions you have encountered in these lessons as an alphabet of movement. You have learned the different letters that you can now begin to put together into words and sentences.

The Feldenkrais Method was created because one person refused to accept what were then considered by the medical field unchangeable limitations. He followed his conviction that we as humans haven't even begun to make full use of our senses, our bodies, our minds; how can we know the extent of what is possible for us?

Today people that would have been confined to wheelchairs walk, others who suffered from debilitating pain now lead active lives, victims of accidents recover faster and more fully, musicians can continue to play their instruments; the Feldenkrais Method® has transformed the lives of countless people.

I founded the Movement Studies Institute and have spent the last 15 years training Feldenkrais practitioners and teaching the method to various health professionals and the general public, to make this information as widely available as possible. many of the lessons in this book are featured in my *Intelligent Body®* tape series, used by thousands of people at home and in wellness centers and clinics. For those with more limitations, we developed the video series *The Timeless Body: Improving with Age*. For people interested in studying the method further to deepen their personal development or to establish a new career, seminars and training programs are available.

These educational resources can make a dramatic impact in your life. It's like having a class for yourself at home. My voice paces you, suggests what to feel and what to sense, and guides you through the movements as if I were there instructing you in person. These lessons are also fun—students have told me they get together for "classes" in their living rooms!

It is my hope that this book and the following educational tools and resources will make a significant difference to you as well. Enjoy doing the lessons, use the audio and visual materials, and discover the transformative power of learning to use your intelligent body.

RECOMMENDED RESOURCES

AUDIO CASSETTE PACKAGES

The Intelligent Body®: Volumes I & II

Improve posture, flexibility, and muscular efficiency while deeply relaxing chronically stiff muscles. Special attention paid to the use of lower back, neck, and balance. Applications have been useful to physical therapists, physicians, sports trainers, and individuals of all ages. A comprehensive introduction to Awareness Through Movement®.

12 audio cassettes | 24 45-minute lessons | User's Manual $130

The Intelligent Body®: Volume III

Following popular demand and the success of Volumes I and II, Dr. Wildman has introduced a third volume to the Intelligent Body series. Volume III provides twelve advanced lessons with exercises specially designed to enhance the listener's understanding of the Feldenkrais Method.® Lesson areas include: posture, rolling, walking, and running.

6 audio cassettes | 12 45-minute lessons $70

VIDEO CASSETTE PACKAGES

The Timeless Body®: Improving with Age

One difference between feeling old and feeling young is how effortlessly we move. This innovative teaching tool is the first and only Feldenkrais-based gerontological/physiotherapeutic video series available in the United States. Three video cassettes and accompanying written manual offer nine complete movement lessons to restore and maintain functional ability, flexibility, and pleasure in the older body in motion.

3 90-minute video cassettes | 9 lessons | User's Manual $159

INDIVIDUAL AUDIO CASSETTES

Dealing with Back Pain

After many years of working with back pain patients, Dr. Wildman developed a unique program at the University of California Back Care Course based on the Feldenkrais Method®. This regimen is helpful both in treating and preventing back pain. This single, 90-minute cassette of easy-to-follow exercises has proven beneficial to back patients as well as to health professionals for use in their clinics. Back pain does not have to be a part of one's daily life.

Audio cassette $15

The TMJ Tape

Dr. Wildman has developed a totally new approach to improving the freedom of the jaw for those with TMJ (temporomandibular joint) problems. Using movements based on the principles of sensory-motor learning, these exercises create greater freedom of motion and reduce muscle stress in the neck and jaw. The exercises consist of: movement of head, neck, and shoulders in relationship to the primary movements of the jaw; relating the tongue, lips, face, and eyes to the jaw; and connections between posture and stress in the jaw. These may also be useful to people with frequently occurring tension headaches.

Audio cassette $15

The Better Driving Tape

Reduce stress, headache, and backache while driving. The Better Driving Tape uses pleasant and simple exercises designed to relax the neck and shoulders, relieve lower back pain, and maintain alertness behind the wheel. Also included are exercises for breathing, posture, and vision.

Audio cassette $15

Moving from Pain into Pleasure: Fibromyalgia and Chronic Pain

This innovative audiotape consists of 5 short Awareness Through Movement® lessons that have proven effective in restoring pleasurable movement for hundreds of people suffering from fibromyalgia and chronic pain.

Audio cassette $15

Mastercard and Visa accepted by mail, phone, or fax; checks by mail. Bulk pricing available. Please allow 2-4 weeks for delivery. For Canadian and overseas orders, call for shipping prices. Prices are subject to change.

For additional information on products, professional training programs, and seminars, or to make an appointment for private Feldenkrais lessons with Dr. Frank Wildman, contact:

The Movement Studies Institute
P.O.Box 2007, Berkeley, CA 94702-0007
Phone (800) 342-3424
Fax (510) 548-4349
www.movementstudies.com

If you want to locate a certified Feldenkrais practitioner near you contact:

Feldenkrais Guild® of North America
3611 SW Hood Avenue, Suite 100
Portland, OR 97201
Phone: (503) 221-6612 or (800) 775-2118
Fax: (503) 221-6616
www.feldenkrais.com